PRECISION POWER

USING AI IN REAL ESTATE DEVELOPMENT

THE EXECUTION INTELLIGENCE BLUEPRINT

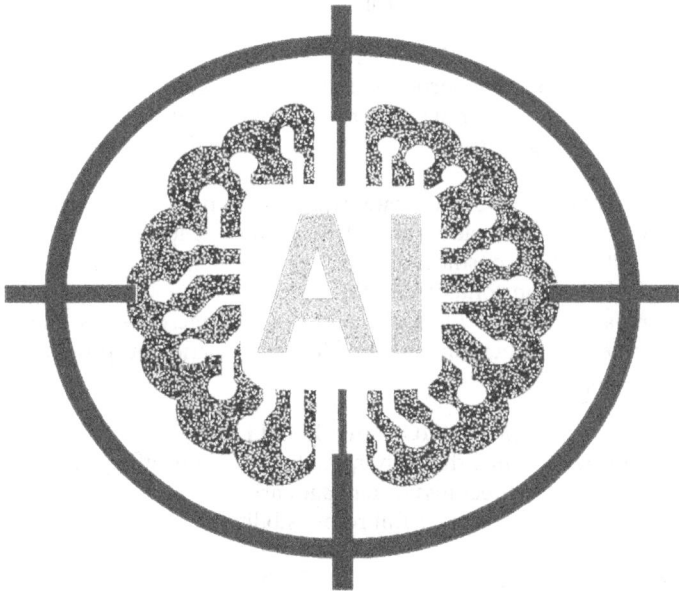

WILLIAM BÖLL

Precision Power Using AI in Real Estate Development: The Execution Intelligence Blueprint
Copyright © 2025 William Böll

This work is published with the understanding that the author and publisher are not engaged in rendering legal, financial, investment, or other professional services. If expert assistance is required, the services of a competent professional should be sought.

Publisher: Velocity Legacy Publishing LLC
ISBN Print: 979-8-9997668-0-9
ISBN eBook: 979-8-9997668-1-6
ISBN Hardback: 979-8-9997668-2-3
Library of Congress Control Number: 2025920945

1. Real Estate Developers 2. Luxury Investor 3. AI-Powered Entrepreneurs
4. Syndicators & Capital Raisers 5. Specialized Niche Builders
6. High-Ticket Sales Professionals 7. Brand-Driven Operators
8. Global Expansion Strategists

I PRECISION POWER. II William Böll

DISCLAIMERS

This book is for educational purposes only. The strategies, frameworks, and examples are based on the author's own real estate experience, entrepreneurial practices, and business philosophies. Nothing in this book constitutes a guarantee of success, earnings, or results.

Success in business and real estate requires judgment, due diligence, skill, and disciplined execution. Readers are solely responsible for the decisions they make and are advised to consult legal, tax, or investment professionals before acting on any information contained in this book.

All company names, trademarks, and product references are the property of their respective owners and are used for identification purposes only. No endorsement is implied.

LIMITATION OF LIABILITY

Under no circumstances shall the author, publisher, or any affiliated parties be liable for any direct or indirect loss or damages arising from the use or misuse of this material. The reader accepts full responsibility for their choices, results, and business outcomes.

For special customized print runs, or Rights or Licensing Agreements: inquiries@velocitylegacypublishing.com
Covers by William Böll
Editor: Mel Cohen
Proofreader: Tracy Johnson
Layout and Design by Megan Leid
Publishing Advisor: Mel Cohen inspiredauthorspress@gmail.com
Printed in the United States of America
Website: www.velocitylegacypublishing.com

TESTIMONIES

I've read hundreds of strategy books, but what you hold in your hands is more than a manual—it's a weapon. It proves what I've believed for years: that execution speed, when combined with precision and narrative, can create disproportionate outcomes. But let me tell you something important: wealth without conscience is fragile.

This book shows how to use artificial intelligence to compress timelines, pre-sell assets, and raise capital with confidence. It's smart, tactical, and unapologetically aggressive. I respect that. Yet if you stop there—if all you see is margin, spread, and replication—you'll miss the bigger picture. Because here's the truth: real estate isn't the real asset. People are.

The most valuable currency in any deal isn't cash flow or comps—it's trust. And trust, once earned, can never be disrupted by technology or copied by competitors. Trust is where legacy is forged. In my work with global leaders through Friends of Peter, I've seen the best systems fall apart when they weren't rooted in conscience. I've also seen fragile startups turn into empires overnight because they were built on relationships, character, and clarity of purpose.

So I encourage you to take what William and JT have built here—and add a layer. The Legacy Layer. Ask yourself:

How can these same AI frameworks accelerate not just wealth, but healing?

How can identity real estate become identity leadership?

How do we ensure that every investor, every partner, every city touched by these projects walks away with something in them, not just for them?

Do that, and you won't just build developments—you'll build movements. That's when specialized real estate stops being a fragile piece of profit and becomes a generational engine of transformation. Because in the end, your legacy is not what you do for people. It's what you leave in people.

—Peter Strople
Founder, Friends of Peter; Creator of Instant Change™

"In a world where most real estate educators talk theory, William Böll delivers execution. Their practical frameworks, branding insights, and monetization strategies have helped us improve how we build, scale, and grow our proptech platforms. They have a unique way of understanding real estate and seeing opportunities ahead of the curve. Working with him has definitely added value to what we do."
—**Harri Majala**
Oulu, Finland
gbuilder.com
3rd generation entrepreneur, 2nd generation homebuilder, 1st generation ConTech founder

AI is reshaping real estate into an arena of predictive power. This book, PRECISION POWER USING AI IN REAL ESTATE DEVELOPMENT, is your telemetry dashboard. It demystifies how machine learning can forecast market shifts in underserved sectors, automate due diligence on quirky assets like adaptive reuse projects, and execute trades with the efficiency of a pit crew under pressure. No longer do you need to guess at tenant turnover in a boutique hospitality flip or valuation anomalies in industrial retrofits; AI hands you the algorithms to spot them, score them, and seize them.

What excites me most about the author's approach is the emphasis on execution. In racing, strategy is worthless without the nerve to commit at 200 miles per hour. Here, you'll find not just theoretical models but battle-tested tactics: from AI-powered scenario modeling that anticipates regulatory headwinds in heritage properties to portfolio optimization tools that balance yield against volatility, much like tuning a suspension for the perfect blend of grip and speed. These strategies aren't for the faint-hearted investor chasing cookie-cutter condos—they're for the specialist who thrives in the turns, where others see only straightaways
— **James Bondurant**
Professional Race Car Driver

PRECISION POWER USING AI IN REAL ESTATE DEVEL-OPMENT is more than a guidebook — it's a blueprint for the future. Within these pages, you'll learn how AI can identify opportunities that would have taken teams of analysts weeks to find, predict market shifts with astonishing precision, and streamline decision-making so you can act when timing matters most.

As someone who spends my life spotlighting innovators and thought leaders, I can tell you: this book is exactly what the industry needs right now. It doesn't just explain the technology; it shows you how to put it to work — profitably, practically, and responsibly.

—Brandon Jay
CEO & FOUNDER | IMA Entertainment Holdings
imaentertainment.com

"Luxury auto condominiums aren't just garages; they're purpose-built environments engineered for performance, security, and elegance. Designed with soaring ceilings, climate-controlled interiors, and architectural finishes, these spaces transform storage into an experience. Every element— from reinforced construction to advanced security systems— reflects a commitment to safeguarding prized collections while creating a setting worthy of them. It's not simply real estate; it's a product that elevates the way automobiles are preserved, displayed, and celebrated."

—Fernando Labastida
Viral Genius Framework
Fernando is a pioneer in Content Marketing and a best-selling author. He is the author of The Startup Book, which explains how to design a new business category and establish yourself as a thought leader. He also hosts the AI Marketing Case Studies podcast and advises CEOs and individuals on how to write their Startup Book.

CONTENTS

ACKNOWLEDGMENTS

I owe a profound debt of gratitude to JT Foxx, whose strategic insights and mentorship were instrumental in shaping both the concepts and execution of this work. JT's expertise in business strategy and his ability to see opportunities where others see obstacles provided the foundation for many of the frameworks presented in these pages. His guidance helped me and my company navigate complex strategic challenges and refine the practical applications that make this book actionable for readers.

The fusion of JT Foxx's CoreBrain.ai, driven by human instinct, strategic dominance, and an execution-first business philosophy, with cutting-edge artificial intelligence created a synergistic force that transcended what either could accomplish alone.

This is not merely knowledge; it is a precise system for thinking, deciding, and executing like a 9-figure entrepreneur. Using this strategy transformed my real estate development company from barely surviving to a high-performing, profitable enterprise.

Any insights of value in these pages are a direct result of this powerful synthesis of human expertise and machine-driven clarity.

— William Böll

FOREWORD

Luxury is not a price point; it's a mindset! And in the world of real estate, nothing embodies that mindset more than the rise of luxury auto condominiums.

No longer just real estate, it's identity real estate. You are not selling square footage. You're selling status, exclusivity, and lifestyle. And in a high-net-worth market where every detail matters, AI is not a buzzword; it's a competitive weapon.

The author of this book gets it. It is not about theorizing what AI might do. It is about implementing what it must do—now. From predictive analytics on high-value car owner demographics to AI-driven site selection, precision marketing, and frictionless customer experience, this book shows you how to use technology to build more than buildings. It shows you how to build obsession-worthy investments.

Luxury auto condominiums are niche; however, the margins are anything but. You are not only dealing with buyers; you're dealing with collectors, chief executive officers, and brand-driven power players. They don't buy what is available. They buy what feels personally built for them. AI allows you to anticipate their behavior and engineer it into every phase of execution.

This book offers what few do: strategy and monetizable execution. It is bold, precise, and unapologetically tactical. If you are in the business of creating desire, not just space, then this is your playbook.

INTRODUCTION

In today's real estate economy, you don't win by being the biggest. You win by being the smartest. This book is not a technical manual. It is a weapon!

The landscape of specialized real estate development; luxury auto condominiums, branded lifestyle communities, medical plazas, niche-use industrial; has never been more competitive, fragmented, or misunderstood. Most developers are trapped in legacy thinking, reliant on intuition, outdated spreadsheets, and bloated teams. They overbuild, overpay, and underperform.

This book rewires that. This book reshapes that mindset.

Inside these pages, you will unlock the JT Foxx AI Execution System: a tactical, field-tested strategy that deploys artificial intelligence not for novelty, but for profit. It is not about "futuristic concepts." It is about immediate control, rapid monetization, and high-leverage execution. You will see how AI, when embedded strategically, eliminates bottlenecks across:

- Site acquisition and zoning manipulation
- Feasibility modeling and investor confidence cycles
- Authority branding and buyer pre-positioning
- Global license replications and scalable exits

You'll learn how to dominate in developments that don't compete on a mass scale, but on margin, velocity, and positioning. You will master the psychology of investor influence, the architecture of trust, and the systems that allow lean teams to control multimillion-dollar portfolios without the traditional drag of bureaucracy or analysis paralysis.

Whether you're a developer, fund manager, syndicator, or builder seeking more efficient capital deployment and stronger exits, this book is your next evolution.

It is not about theory. It is about execution intelligence. You'll get:

- AI-driven frameworks for due diligence, model optimization, and deal stacking
- Authority-building content strategies that attract higher-net-worth investors
- Automation templates that compress your development cycle from months to weeks
- Brand leverage tactics that elevate your positioning without increasing costs

This book is not for spectators. It is for strategic operators ready to weaponize AI to out-think, out-build, and out-profit their competition.

If you're ready to build data-driven, investor-magnetic, globally scalable developments, this is your unfair advantage.

PREFACE

It is highly recommended that the reader obtain a copy these two JT Foxx bestselling books *Business Is War: If You Want To Win, Learn From Failures, Not Success* and *Business is War: AI is the New Weapon*. They provide foundational insights into the mindset, strategies, and tools required to thrive in today's competitive environment.

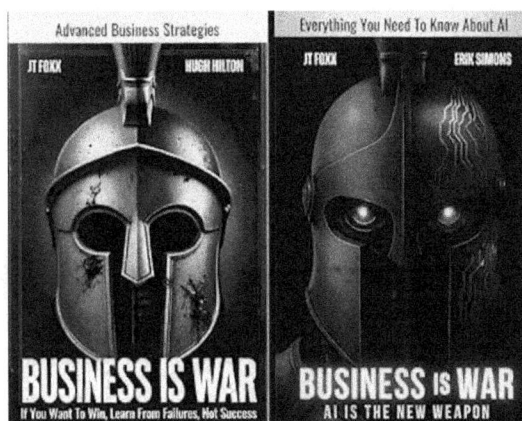

The first volume emphasizes the value of lessons drawn from setbacks—reminding entrepreneurs and leaders that resilience and adaptation often create more lasting strength than success alone. The second book advances this philosophy into the modern era, positioning artificial intelligence as the decisive advantage in business strategy, growth, and execution.

At Velocity Performance Alliance, these principles have had a direct and transformative impact. JT Foxx's frameworks have helped shape our business strategy and operations, ensuring that our organization remains disciplined, forward-thinking, and aligned with the most effective tools for sustainable growth.

Together, these works serve as both a compass and a battlefield manual for those determined not just to compete but to dominate in business. Readers who study them will gain an invaluable foundation for understanding the principles of strategic positioning, innovative thinking, and disciplined execution that underpin lasting success.

CHAPTER 1

WHY AUTO CONDOMINIUMS ARE THE NEXT TROPHY ASSET

*"Most people chase trends.
Smart entrepreneurs engineer them." – JT Foxx*

*"Niche development isn't about filling space—
it's about solving high-value problems that the
market hasn't articulated yet." – William Böll*

We don't develop property. We develop categories. You are not building real estate. You're engineering emotional demand. You are not a developer. You're a category owner. And when you own the category, you don't ask for market value. You set it.

Auto condominiums are not a passing trend. They are the next trophy asset class, the perfect fusion of lifestyle branding, scarcity economics, and investor psychology. They are engineered, not for the mass market, but for the passion-driven elite. These people don't flinch at premium pricing because they're not buying space. They're buying identity, belonging, and status.

When you own the niche, you own the margins. That's Velocity Real Estate 101.

This chapter breaks down why auto condominiums are the next blue-ocean opportunity and how to execute with AI, positioning, and speed.

1. YOU ARE NOT SELLING STORAGE. YOU ARE SELLING STATUS.

- Auto condominiums are not garages. They are identity sanctuaries. They are designed for ultra-high-net-worth (UHNW) collectors who don't want to store cars; they want to showcase legacy.
- These are private galleries. Lifestyle vaults. Trophy rooms. You're not marketing to car owners. You're speaking to status buyers who trade on image, not logic.
- They don't negotiate price; they negotiate exclusivity.

2. SCARCITY = LEVERAGE

- Auto condominiums are intentionally limited. That's not a bug; it's a feature.
- Zoned for industrial/showroom use (rarely rezoned for residential)
- Customizable and high-finish by default
- Difficult to replicate due to community pushback, design complexity, or land scarcity
- This constraint creates demand tension—what JT calls "strategic bottlenecking." And when supply is intentionally capped, you don't chase comps; comps chase you.

3. THE DEMAND SURGE IS HERE. SUPPLY IS NOT.

Auto condominiums sit at the intersection of three surging macro trends:

- Explosive growth in the collector car market (post-pandemic spike in tangible asset investing)
- Global rise in high-net-worth (HNW) buyers who collect cars, watches, and alternative assets
- Escalation of demand for private, secure, climate-controlled, prestige-based storage

Meanwhile, developers are way behind. Most have no clue how to:
- Zone properly
- Price emotionally
- Brand with elite positioning

You will now have an execution gap advantage. Use it.

4. LOW OVERHEAD. HIGH MARGIN. NO RESIDENTIAL HEADACHES.

Auto condominiums are structurally simple:
- Steel-frame warehouse + luxury interior finishes
- No toilets, tenants, or turnover risk
- No hospitality licenses, rent control, or short-term rental (STR) bans

They generate:
- Homeowners association (HOA) fees
- Event revenue
- Club memberships
- Concierge service upsells

High-touch. Low-staff. Big spread. That's profitable simplicity.

5. EMOTION = PRICE ELASTICITY

Real estate built for emotional buyers has one rule: value is felt, not calculated.

Auto condo buyers will:
- Spend six figures on interior upgrades
- Host private events to show them off
- Brag about ownership because it signals status

Cap rates are irrelevant. Emotion is the currency.

6. EXPORTABLE GLOBAL CONCEPT

It is an international brand asset, not a local play.

From:
- Naples to Newport Beach
- Scottsdale to Singapore
- Melbourne to Monaco
- Wherever there are high-net-worth individuals (HNWIs) and luxury vehicles, there is a need for this product.

Build once. Franchise the brand. License the model.

7. TROPHY BEHAVIOR, BOUTIQUE SCALE

Most successful developments range from 12 to 30 units. Why?
- Easier capital raise
- Controlled execution
- Faster absorption
- Phased scalability

All while commanding trophy pricing. It is institutional upside with entrepreneurial speed.

8. REAL ESTATE IS THE PLATFORM. THE BRAND IS THE MOAT.

The top developers aren't selling square footage. They're selling membership in a brand.

Add layers:
- Car collector clubs
- Private rally events
- Dealership affiliations
- Custom design partnerships
- Media and influencer tie-ins

The result? Lifestyle monetization. Real estate becomes the entry point, not the endpoint.

9. VERTICAL + HORIZONTAL MONETIZATION

Multiple monetization layers = insulation + scale.

Vertical:
- Sell units to users or investors
- Retain management, concierge, or design services

Horizontal:
- Event rentals
- Sponsorships
- Private client upgrades
- Brand licensing

You will learn how small-footprint projects yield outsized returns.

10. DEVELOPER ADVANTAGE: LOW RED TAPE, FAST CYCLE

Most auto condo projects bypass typical residential headaches.
- Faster permitting
- Light industrial or commercial zoning paths
- Streamlined construction timelines
- Lower density = less community resistance

Build lean. Launch fast. Scale on brand.

11. YOU'RE RIDING A MACRO SURGE IN WEALTH AND COLLECTIBLES

Wealth is growing. Collectible car ownership is exploding.
- Over 65% of UHNWs own at least one collectible vehicle
- Post-pandemic behavioral shift favors privacy, mobility, and physical assets
- Younger HNW buyers are more image-conscious and lifestyle-driven

The smart money goes to assets that blend identity, security, and upside.

12. EXECUTION STRATEGY

Land Strategy:

Acquire parcels near private airports, country clubs, or affluent suburbs: fringe luxury locations with identity equity.

Pre-Sell Model:
- Use AI and psychographics to segment buyer profiles
- Launch tiered pricing: Founders Circle → Premier → Late Access

- Frame scarcity: "Only 18 units. Never repeated. Ever."

Brand Positioning:
- Package as a lifestyle investment
- Elevate design, digital presence, and buyer onboarding to reflect premium exclusivity

Community-Building:
- Host private events before completion
- Introduce buyers to each other early
- Build story equity around ownership

BOTTOM LINE:

Auto condominiums are not a product. They're a movement driven by identity, exclusivity, and asset protection. This is where passion for high-performance vehicles meets real estate as a luxury investment class. It is no coincidence that world-renowned collectors like Jay Leno, Ralph Lauren, and Sheikh Hamad bin Hamdan Al Nahyan have invested in purpose-built automotive spaces that are as curated as the cars themselves. For elite buyers, it is not about storage—it's about status, legacy, and leverage. The auto condo movement is the monetization of culture, and those who understand it early won't simply participate—they'll dominate.

It is where emotional capital meets financial capital and where developers who understand both can engineer generational assets.

If you position it right, you don't compete on comps. You control the narrative, command the premium, and create a forever asset.

Successful real estate developers approach it this way. Now make it your advantage.

CHAPTER 2

THE AI ADVANTAGE IN SPECIALIZED DEVELOPMENTS

How artificial intelligence unlocks
precision, speed, and investor
confidence in niche real estate ventures

"Technology doesn't replace you. It exposes whether you were needed in the first place." – JT Foxx

"AI doesn't replace fundamentals—it scales them. The edge comes when discipline meets data." – William Böll

Artificial intelligence unlocks precision, speed, and investor confidence in niche real estate ventures. And in specialized developments like auto condominiums lies precision execution, buyer intelligence, and scalable monetization. Here's how we strategically deploy AI to dominate this vertical:

1. HYPER-PRECISION SITE SELECTION
AI integrates geospatial data, demographic trends, income clustering, psychographics, and traffic flows to pinpoint ideal sites before they hit the open market.

Example: For a luxury auto condominium, AI can cross-reference UHNW car collector density with underutilized industrial-zoned parcels and forecast appreciation zones ahead of local brokers.

2. ZONING & ENTITLEMENT STRATEGY
AI can rapidly analyze local zoning codes, planning commission minutes, and permit history to:
- Identify the path of least resistance

- Predict public opposition
- Recommend optimal use-case narratives

This reduces legal overhead and fast-tracks entitlements.

3. DESIGN OPTIMIZATION

AI-powered generative design tools create and test thousands of unit configurations, parking layouts, and visibility angles in seconds. You get:

- Maximum price per square foot
- Optimal layout for light, traffic flow, and luxury appeal
- Cost-effective construction sequencing

4. CAPITAL STACK ENGINEERING

AI can simulate investor return waterfalls, risk-adjusted capital stack structures, and tokenized equity models, optimizing internal rate of return (IRR) for Limited Partners (LPs) while protecting General Partners (GP) downside.

Bonus: Use AI to run thousands of simulations to find the perfect blend of debt, preferred equity, and common shares that make your project bankable and attractive to investors.

5. BEHAVIORAL BUYER TARGETING

Instead of mass-market ads, AI analyzes social graph data, past purchase behavior, and micro-signals to target qualified emotional buyers.

Use case: Find every Ferrari owner in Southern California who also follows architecture influencers and collect leads before they raise their hand.

6. AI-AUGMENTED SALES SCRIPTS

Use conversational AI (trained on top closers) to build custom pitch scripts, live chatbots, or video AI closers tailored to niche segments.

Example: A buyer obsessed with racing history gets a different AI-led sales path than a collector who owns rare Bentleys.

7. INVESTOR CONFIDENCE ENGINE

AI dashboards forecast construction timelines, cost overruns, lease-up velocity, and resale comps in real time, keeping investors updated and confident.

Turn uncertainty into clarity and quarterly reports into predictive insight.

8. FRANCHISE OR LICENSING REPLICATION

Once the AI learns what works (design, targeting, ops), you can replicate your specialized development model across markets at speed and precision.

AI becomes your expansion chief operating officer (COO).

9. HYPER-TARGETED BUYER PROFILING (AI-DRIVEN AVATAR CREATION)

JT teaches that you don't sell to everyone; you sell to someone specific. AI can help you build deep psychographic and behavioral profiles of ideal buyers (exotic car collectors, luxury recreational vehicle (RV) owners, and high-net-worth investors).

Use AI to:
- Track digital signals (searches, social behavior, and wealth indicators)
- Create segmented ad audiences with buyer-specific language
- Predict who's not just interested but ready to buy

Result: You save **80%** of your marketing budget and double your conversions.

10. SMART PRICING MODELS (DYNAMIC & TIERED)

AI can analyze buyer behavior, market comps, and scarcity curves to set tiered pricing models that adjust in real-time based on interest, urgency, or exclusivity.

Use it to:
- Create "Founder Tiers," "Platinum Garage Rows," and dynamic premiums

- Simulate the impact of scarcity and price anchoring
- Detect undervalued upsell opportunities (concierge, detail bays, and club access)

Real-time monetization intelligence: our strategy of extracting every ounce of value.

11. VIRTUAL DESIGN & CUSTOMIZATION TOOLS

Auto condominiums are luxury real estate. The more bespoke they feel, the more people pay. AI enables:

- Virtual unit customization (floors, lifts, and lounges)
- Real-time visualization + cost calculation
- AI-styled finish suggestions based on client taste or budget

You sell the dream before the concrete is even poured.

12. OPERATIONAL AUTOMATION (AI-POWERED BACK END)

Speed and systemization scale margin. AI lets you:

- Automate lead follow-up, contracts, and onboarding
- Power virtual concierges for owners
- Optimize unit scheduling and resale strategies

Lower overhead, faster delivery, and higher margin become your real estate trifecta.

13. PREDICTIVE INVESTOR BEHAVIOR & EXIT MAPPING

AI helps forecast who's likely to flip their unit, who will hold, and who will become your next capital partner. You can:

- Identify early resale churners and lock them into equity-sharing
- Predict which cities or verticals are next (luxury storage for art, wine, and watches)
- Align resale strategy with your next fundraise or brand expansion

14. EXIT STRATEGY INTELLIGENCE

AI models potential buyers (real estate investment trusts (REITs), family offices, international funds), predicts market timing, and crafts exit narratives based on historical acquisition behavior, not guesswork.

You sell the brand platform and data flywheel, not just the real estate.

KEY TAKEAWAY:

AI gives you an edge at scale, understanding your buyer before they speak, optimizing every dollar, and systematizing every step from marketing to monetization.

CHAPTER 3

FRAMING CAPITAL RAISES WITH AI + EMOTIONAL EQUITY

Raising capital is not a numbers game.
It is a framing game.

"You don't wait for the right opportunity—you build it with data, speed, and audacity." – JT Foxx

"You don't chase markets. You interrogate data until it reveals demand others can't see." – William Böll

In today's investment climate, spreadsheets don't move money, stories do. Capital flows to perceived certainty, not technical perfection. That's why most developers lose. They pitch logic while elite capital chases identity, trust, and clarity.

When you combine emotional framing, positioning authority, and AI-enhanced investor targeting, you create capital momentum that compounds. You raise faster, you raise smarter, and you raise on your terms.

This chapter delivers the full methodology on how to raise investor capital using AI-enhanced systems, persuasive psychology, and bulletproof framing.

1. YOU'RE NOT RAISING MONEY. YOU'RE SELLING A FUTURE.

Most developers try to raise capital by pitching:

"Here's the IRR. Here's the comp set. Here's the pro forma." Wrong game. That's analysis. Investors buy narrative, not spreadsheets.

Your job is to shift from pitching the project to selling the inevitability of its success. That comes down to:

- Framing yourself as the certainty in the deal
- Positioning the investor as a visionary, not just a financier
- Anchoring the conversation in emotion, not math

AI plays the supporting role. Framing wins the deal.

2. THE JT FOXX FRAMING STACK FOR CAPITAL RAISES

Here's the execution structure:

Step 1: Anchor With Certainty

Open with a conviction statement:

"This is not just another development. It is a trophy asset engineered to attract irrational buyers, not logical ones. That's how we'll win."

Why this works:

- Confidence transfers certainty
- Dismisses objections before they form
- Moves the investor from evaluation to alignment

Step 2: Elevate Their Identity

Position the investor as elite:

"We're only bringing this to five investors who are aligned with lifestyle, legacy, and brand equity. Not just returns."

Repositions as:

- The deal is exclusive
- The investor is rare
- The capital raise is selective, not needy

Step 3: Stack the Emotion + Logic

You don't avoid the numbers; you sequence them after trust and story. Lead with:

- Market trends they are already aware of
- Market trends they already agree with
- Scarcity economics (limited land, rare zoning, and rising car culture)

- Emotional buyer behavior that drives pricing

Only then show:
- IRR potential
- Exit plan options
- Capital expenditures (CapEx) and operating expenditures (OpEx) detail

3. AI TOOLS THAT AMPLIFY INVESTOR VELOCITY

AI doesn't replace your pitch. It multiplies your momentum.

AI Use Case #1: Investor Avatar Targeting

Tool Examples: SparkToro, Facebook Audience Insights, ChatGPT
- Profile existing auto condo buyers
- Build segmented investor avatars: legacy builders, prestige seekers, and return on investment (ROI) syndicators

Output:
- Precision-targeted ad campaigns
- Email copy tuned to identity, not just IRR

AI Use Case #2: Emotional-Logic Deck Builder

Tool Examples: Beautiful.ai + ChatGPT
- Prompt: "Create a 12-slide investor deck that combines narrative framing, scarcity leverage, and IRR logic."

Slides include:
- Trophy asset psychology
- Emotional market insights (collector car surge, generational wealth shift)

Deal execution certainty: "Why we win even if comps don't exist."

AI Use Case #3: Pre-Call Investor Intel

Tool Examples: Crystal + Clay + LinkedIn AI Extensions
Before investor meetings:
- Use AI to analyze LinkedIn profile tone
- Match pitch language style to decision-maker personality (e.g., visionary versus data analyst)

Result:
- Adaptive pitch delivery
- Higher conversion in fewer meetings

4. FRAMING THE OFFER LIKE A PREMIUM ASSET

Never pitch with generic offers. Use tiered positioning with time scarcity and founder psychology:

Tiered Capital Stack Example:

TIER	AMOUNT	INCENTIVE	FRAMING
Founder	$250K+	5% GP equity + naming rights	"Legacy-level investor. Branded into the build."
Premier	$100K–$249K	Priority unit selection	"First-mover access. Early equity alignment."
Core	$50K–$99K	Clubhouse membership + concierge credit	"Lifestyle return + hard asset security"

AI can auto-generate emails, videos, and chat sequences for each tier using conditional logic.

5. AUTOMATED NURTURE = CAPITAL MOMENTUM

Raise once? No. Build an AI-driven investor funnel that never sleeps.

AI Investor Nurture Stack:
- Customer relationship management (CRM): High-Level or HubSpot
- Email/Video: AI-written drip sequences that educate + elevate

- Chatbot: Investor Q&A automation
- Webinar: AI-assisted live or evergreen pitch event
- Follow-Up: Smart reminders, scarcity framing, deadline triggers

Emotional proximity = faster capital conversion.

6. INVESTOR CLOSING SCRIPTS (FRAMING OVER FORCE)

"What if the market cools?"

"We're building for trophy buyers, not speculators. Emotionally-driven markets never price the same way as logical ones. That's our edge."

"What's my downside?"

"You're buying into a branded asset class with limited supply, global demand, and multiple exit paths. Downside is managed through margin, not hope."

"Why now?"

"Every month you wait, scarcity increases and the entry point rises. Founders don't follow; founders own."

BOTTOM LINE:

You're not pitching real estate. You're offering emotional alignment + elite access.

And when you frame the capital raise like a premium identity product with AI support and emotional storytelling, you shift from chasing capital to commanding it.

Chapter 4

AI-DRIVEN SITE SELECTION & ZONING DOMINATION

Winning the land war before it starts

"Assumptions kill deals. Precision makes you rich."
– JT Foxx

"A profitable project doesn't begin with enthusiasm—
it begins with an unflinching look at risk and return."
– William Böll

In real estate, the game is won before the deal hits the market. If you're competing for land, you've already lost. The best developers never chase listings. They engineer off-market certainty, political favor, and hyper-local intelligence before anyone else is even aware there's a play.

And with AI, what used to take months of scouting, zoning research, and political maneuvering can now be compressed into days with far greater precision, less risk, and significantly higher upside.

This chapter unpacks the system for dominating site selection and zoning approval using AI-powered intelligence, negotiation leverage, and brand equity framing.

1. YOU'RE NOT BUYING LAND. YOU'RE BUYING CONTROL.

Land doesn't make you money. Leverage does. You don't win because of location; you win because of the margin between perception and reality. That's where AI gives you the edge.

THE FORMULA:
"Undervalued or overlooked zoning + strategic visibility + emotional upside = exponential ROI"
Most developers buy for potential. We buy for positioning.

2. OUR PROVEN AI SITE SELECTION MODEL
"Find land that looks invisible now but will be untouchable tomorrow."

Step 1: AI Heat Mapping
Tool Examples: Regrid, LandVision, Google Earth Engine

- Overlay income demographics, collector car density, and growth corridors
- Cross-reference: proximity to private airports, marina districts, luxury dealerships
- Use predictive traffic modeling to identify future volume zones

Result:
Find where future value will spike, and buy it before the public sees it.

Step 2: Zoning Flexibility Scan
Tool Examples: Zonar.AI, local geographic information system (GIS) databases

- Detect parcels already zoned for light industrial/showroom (I-1, M-1, C-2)
- Identify municipalities with overlay districts that allow flexible uses
- Run AI-powered zoning change probability models

Key Trigger:
Look for areas with recent rezones in adjacent parcels, which signals political openness.

Step 3: Political Influence Mapping
Tool Examples: GovAI (public records scraper) + ChatGPT-4o

- Map relationships between landowners, planning commissioners, and council members

- Use AI to analyze past rezoning approvals and who sponsored them
- Predict which officials to build early relationships with

Zoning is not about law. It is about influence and precedent.

3. FRAMING THE ZONING PITCH LIKE A BRAND UPGRADE

Municipalities aren't looking for buildings. They're looking for image control. You don't pitch your project as a development project. You frame it as a brand asset for the city.

Zoning Pitch Framework:

Old Frame: "Please rezone, as it is good for your tax base."

New Frame (Our Style): "This project is a regional destination. It attracts affluent collectors, hosts international events, and enhances the prestige of your municipality, all with zero strain on schools, roads, or infrastructure."

Use:

- Renderings of event-ready auto clubs
- Economic impact models
- Public-private win/win frameworks

AI helps simulate economic impact, jobs created, and future tax inflows, then wraps that data in compelling visualizations.

4. OFF-MARKET ACQUISITION STRATEGY: AI + NEGOTIATION

The best parcels are never listed. They are either:

- Owned by "old money" families
- Sitting with land bankers
- Held by small industrial operators with no succession plan

Your AI-Led Approach:

1. Scrape county records for land held 15+ years, out-of-state owners, or vacant industrial

2. Use AI to write personalized outreach letters that trigger curiosity and status. Example: "Your land could be transformed into a regional trophy asset."
3. Leverage social engineering:
   - Use ChatGPT to script custom introduction messages
   - Engage sellers emotionally first, then bring the offer

Negotiation Framing Tip (JT Principle):
"Don't offer a number. Offer a vision and a way to be part of it."

5. WHEN ZONING PUSHBACK COMES: GO OFFENSIVE

Zoning resistance is not a wall. It is a negotiation cue. You win with:
   - Third-party validators (bring luxury car dealers, auction houses, or events as soft endorsements)
   - Community givebacks (AI-modeled donation proposals, public usage incentives)
   - Precedent flipping ("You approved this last year. This has a lower impact.")

Create political air cover. And if needed, use AI to simulate alternate plans (storage, commercial, etc.) to create urgency:
"You can approve our iconic project, or we'll legally build a generic warehouse. Your call."

6. RED FLAG FILTERS FOR SITE SELECTION (AI SCORED)

Use AI to flag deals with invisible drag:
   - Environmental risk (flood, hazmat, brownfield)
   - Community opposition triggers (adjacent schools, HOA pushback zones)
   - Infrastructure gaps (lack of sewer, road access)

Assign AI risk scores to each parcel. Focus only on high-margin, low-friction zones.

7. USE ZONING AS A BRANDING WEAPON

Once zoning is locked, turn the permit into a press event.

- Use AI to generate branded articles: "XYZ City Approves Exclusive Auto Sanctuary"
- Public relations (PR) syndication using AI-written media releases to collector sites, local media, and LinkedIn
- Frame the development as an achievement, not just a project

This builds pre-sell momentum, raises perceived value, and creates early investor fear of missing out (FOMO).

BOTTOM LINE:

Real estate is war. AI is your reconnaissance. Zoning is your weapon.

You don't wait for opportunities; you engineer them with precision, persuasion, and political pre-alignment.

If you dominate site selection and zoning, everything after that is execution.

CHAPTER 5

THE REAL ESTATE WEALTH TRIANGLE

A framework to build generational wealth
faster, using niche real estate plays like
luxury auto condos.

"Real estate isn't about location—it's about timing, leverage, and knowing what others overlook." – JT Foxx

"We don't buy land—we buy positioning. The dirt has to make sense before the design ever can." – William Böll

WHAT IS THE REAL ESTATE WEALTH TRIANGLE?

A strategic model that aligns active income, passive cash flow, and asset appreciation into a compounding wealth engine all in one specialized real estate venture.

When applied to auto condominium development, this triangle becomes even more powerful due to the project's ability to monetize passion, scarcity, and brand.

The Real Estate Wealth Triangle is built on three core pillars:
- Cash flow
- Equity and wealth creation
- Speed to scale

In the context of auto condominium developments, here's how each leg of the triangle comes into play and how to execute with precision:

1. CASH FLOW: SELL, LEASE, OR HYBRID

Our philosophy: Cash flow is the lifeblood. Without it, you're speculating, not investing.

Application to Auto condominiums:

- Pre-Sales: Lock in buyers before construction to fund build cost with deposits, which will minimize or eliminate debt
- Long-Term Leasing: Offer managed leaseback options to owners, generating recurring revenue
- Hybrid Model: Keep premium units as income-producing assets while selling mid-tier units to fund operations

Once operational, the asset throws off passive income through:

- HOA dues and facility fees
- Event rentals and filming
- Luxury storage and detailing services
- Exclusive club memberships
- Optional leased units for absentee owners

2. ASSET APPRECIATION – BRAND, REAL ESTATE, AND EXIT PREMIUMS

Over time, your asset appreciates in three compounding layers:

- Land and structure appreciation in constrained zones
- Brand intellectual property (IP) appreciation because your club becomes a name worth buying
- Exit multiple expansion via REIT buyout, portfolio sale, or joint venture (JV) licensing

Key Insight: Sell more than dirt. Sell the dream, the data, and the designs. Then watch your exit valuation soar.

3. TRIANGLE SYNERGY:

When you lock all three sides (active, passive, and appreciation) you don't just make money. You build a replicable, investor-attractive machine.

Auto condominiums are not a product. They're a triangle-shaped wealth platform.

Key Insight: Cash flow turns this from a flip into a legacy platform. Upsell concierge access like a hotel but with Ferrari keys.

Execution Tip: Use concierge services, climate-control premiums, and branded events to maximize per-unit cash yield.

Key Notes:

- Synergy: Each side feeds the others. Pre-sales validate comps → cash flow funds brand → brand drives asset appreciation.
- Trophy Asset Premium: Once the triangle stabilizes, you can 10x valuation by packaging the brand, data, and operating model into franchisable IP or a portfolio roll-up.

4. EQUITY/WEALTH CREATION – APPRECIATION THROUGH SCARCITY + BRAND

Our model is clear: Own the dirt, control the brand, manufacture the appreciation.

Application to Auto Condominiums:

- Land Arbitrage: Buy undervalued commercial land near HNW hubs, airports, or marina zones.
- Example of the Brand as an Asset: "The Vault by [Your Name]" becomes a brand, not just a building. That brand equity makes each new project more valuable.
- Tiered Ownership and Club Structure: Add equity through exclusivity. Introduce Founders' Rounds, resale restrictions, and buy-back options.

Execution Tip: Offer resale management to extract fees and control pricing and protect the brand premium on secondary sales.

5. SPEED TO SCALE – SYSTEMS, SYNDICATION & STRATEGIC CAPITAL

You don't build wealth one deal at a time; you build it with repeatable systems and scalable capital.

Application to Auto Condominiums:

- Cookie-Cutter Model: Create a repeatable development blueprint of unit mix, build specs, and services you can roll out in multiple cities.
- Investor Syndication: Bring in capital through your personal brand authority. Position as a low-risk, lifestyle-aligned asset.
- License or Franchise Model: Once brand and systems are proven, scale through partners with your oversight, taking a percentage cut.

Key Takeaway: Build your team once. Then scale the machine, not the hustle.

CHAPTER 6

AI BRANDING STRATEGIES THAT BUILD PRE-SOLD DEVELOPMENT

How to engineer perceived value before you
ever break ground

*"A project that doesn't emotionally convert on paper will
never convert in the market." – JT Foxx*

*"Design is only powerful when it aligns with monetizable
outcomes. Anything else is architecture without ROI."
– William Böll*

Most developers think branding is optional. That's why
most developers sell late, sell slow, and sell cheap.

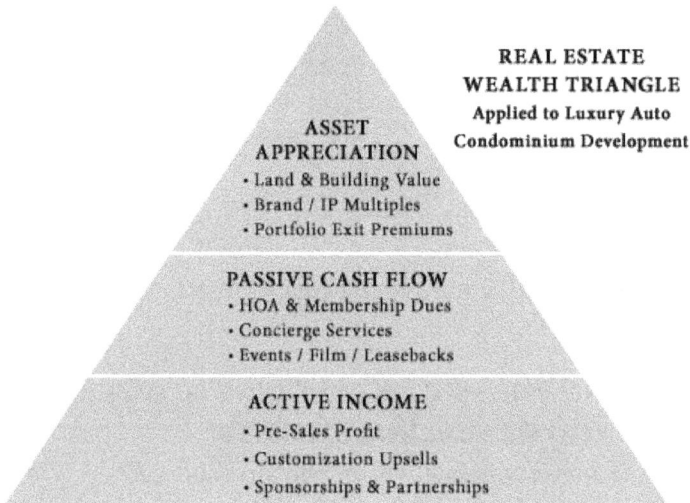

**REAL ESTATE
WEALTH TRIANGLE**
Applied to Luxury Auto
Condominium Development

**ASSET
APPRECIATION**
• Land & Building Value
• Brand / IP Multiples
• Portfolio Exit Premiums

PASSIVE CASH FLOW
• HOA & Membership Dues
• Concierge Services
• Events / Film / Leasebacks

ACTIVE INCOME
• Pre-Sales Profit
• Customization Upsells
• Sponsorships & Partnerships

BRAND FIRST, BUILD SECOND.

The most profitable developments in the world are not the best built; they are the best framed. They are sold out before concrete is poured. And their value is not in the square footage. It's in the status perception engineered around the project.

Branding is not a logo. It is a weapon.

In this chapter, you'll learn our playbook for using AI to brand your development like a luxury product and how to pre-sell with elite positioning before ground is even broken.

1. PRE-SOLD MEANS PRE-BELIEVED

- Your job is not to convince people your project is good.
- Your job is to make them believe they can't afford to miss it.

Branding is not design; it's narrative control. And narrative, when powered by AI, becomes a scalable influence.

2. OUR SUCCESSFUL 5-LEVEL BRANDING FRAMEWORK FOR REAL ESTATE

LEVEL	DESCRIPTION	OUTCOME
1	Name + Logo	Recognition
2	Brand Story	Connection
3	Authority Positioning	Elevation
4	Exclusivity Signals	Scarcity
5	Emotional Identity	Obsession

Most developers stop at Level 2.

We build to Level 5, then we use AI to distribute and scale that identity across platforms, markets, and investor pools.

3. NAMING THE PROJECT: YOU'RE NOT SELLING UNITS. YOU'RE SELLING IDENTITY.

Forget "Parkview Estates" and "Motor Vault 27."

Use AI + luxury psychographics to name your development like a luxury watch, not a real estate commodity.

Use this prompt in ChatGPT:

"Generate 15 brand name ideas for a luxury auto condominium development targeting UHNW car collectors who value privacy, status, and design. The name should evoke legacy, rarity, and elite membership, not garages."

Once you land on a brand name, run:

↑

ASSET APPRECIATION
- Land & Building Value
- Brand / IP Multiples
- Portfolio Exit Premiums

←————————|————————→

PASSIVE CASH FLOW
- HOA & Membership Dues
- Concierge Services
- Events / Film / Leasebacks

↓

ACTIVE INCOME
- Pre-Sales Profit
- Customization Upsells
- Sponsorships & Partnerships

- Domain availability
- Trademark screening

Check the semantic score (how it sounds when said aloud).

AI gives you clarity at speed. You only launch once. Nail the identity.

4. AUTHORITY BRANDING = PRE-SELL POWER

Before construction begins, your brand must look like:
- A proven, global concept
- A safe investment for status-driven buyers

- A lifestyle club, not a speculative asset

Use AI to:
- Script your founder story (hero's journey + elite positioning)
- Build an origin video: 90 seconds, emotion first, logic second
- Auto-generate 12 weeks of content for social/email/web on the concept, market trends, buyer profiles, and collector culture

"People don't buy square footage. They buy belief."

5. VISUAL IDENTITY = PERCEIVED VALUE MULTIPLIER

AI tools like Midjourney, DALL·E, and Runway ML allow you to build:
- Hyper-realistic architectural renders
- Lifestyle imagery that includes luxury cars, events, and owners
- Branded visual packs for investors and social media

Before the land is cleared, the story must be visible.

Use visual AI to:
- Showcase the interior as a gallery, not a garage
- Embed collectibles, cigars, fine art, and racing simulators
- Frame every render with owner psychology: legacy, access, privacy

6. THE PRE-SELL MEDIA STACK (AI-DRIVEN)

Here's the AI-driven content engine you launch before ground breaks:
- Automate the distribution of emotional brand identity at scale.

7. FRAMING SCARCITY USING BRAND SIGNALS

Scarcity drives urgency, but perceived scarcity drives price elasticity.

Use AI-enhanced branding to:

- Limit availability with tiered access: "Founders Phase," "Invite-Only Preview," "Final Tier"
- Embed timers, countdowns, and sold-out signals into web funnels
- Build waitlists using AI chatbots that segment buyers by priority level

Position every communication with this tone:

"This is not available to everyone. And once it is, it won't be for long."

8. LIFESTYLE BRANDING = PREMIUM JUSTIFICATION

Every photo, video, article, and social touchpoint must reinforce one idea:

"This is not storage. It is a sanctuary for the elite."

Embed:

- Track-day event teasers
- Private car reveal parties
- Brand ambassador testimonials
- Influencer walk-throughs
- AI-simulated footage of garage interiors with $30M in vehicles

Emotion justifies premium pricing. AI makes that emotion visual, scalable, and immediate.

9. BRAND EQUITY = LICENSING POWER

When your development becomes a brand, you can:

- License the concept globally
- Sell out faster in future phases
- Create JV demand from dealers, clubs, and influencers

Think like Ferrari. The building is not the asset. The brand is.

Use AI to build the content, the narrative, and the cultural imprint that scales, without hiring an entire agency team.

BOTTOM LINE:

Most developers build, then brand. Successful developers brand, then build.

That is why we pre-sell at premium pricing while others negotiate at a discount. That is why our buyers don't compare; they commit.

And now, with AI, you can build the identity engine that drives perceived value before a single unit exists.

CHAPTER 7

AI IN SITE SELECTION AND ZONING BATTLES

*"If you don't understand the money,
you'll always work for someone who does." – JT Foxx*

*"Structure equals survivability.
Profitability is only as strong as the capital stack
that underpins it." – William Böll*

Strategy & Development using AI refers to the integration of artificial intelligence across every phase of planning, designing, financing, and executing a business or real estate project. It enhances human decision-making with data-driven speed, pattern recognition, and predictive precision.

AI turns intuition into insight, blueprints into simulations, and guesses into executable strategies at scale. Using AI in site selection and zoning battles is not optional anymore; it's how elite developers create unfair advantages. You should approach this with strategic precision, combining speed, data, and positioning to outmaneuver competitors and regulators.

Here's the playbook:

1. STRATEGIC PLANNING

AI is used to:
- Analyze market gaps, pricing models, and competitor weaknesses
- Identify underserved customer segments or geographic zones

- Simulate hundreds of scenarios based on macroeconomic variables

Key Takeaway: Instead of spending months in spreadsheets and war rooms, AI gives you a data-validated roadmap in hours.

2. AI IN SITE SELECTION; FIND THE GOLD BEFORE OTHERS DO

Key Principle: Buy what is not obvious now. Make it obvious later.

AI Execution:

- Heatmap Algorithms: Use AI to scan and rank properties based on proximity to high-net-worth zip codes, traffic patterns, airports, car dealerships, income levels, and future infrastructure plans.
- Predictive Modeling: Feed in historical appreciation data, zoning shifts, and buyer trends to predict future high-demand areas.
- Parcel Optimization: AI can identify underutilized or miss-zoned parcels and score them for highest-and-best-use (e.g., converting light industrial to luxury storage).

Key Takeaway: Developers used to rely on brokers; now they rely on trained algorithms that see invisible patterns in millions of data points.

2. AI IN ZONING BATTLES – WIN THE GAME BEFORE IT STARTS

Key Principle: Control the narrative. Control the outcome.

AI Execution:

- Sentiment Analysis on Public Records & Social Media: Gauge neighborhood support or resistance before you file. Identify potential allies and opponents.
- Zoning Approval Likelihood Modeling: AI compares your project to thousands of previous applications

in that jurisdiction, estimating approval probability and key objection triggers.

- Smart Proposal Engineering: Use AI to craft proposals that preemptively address opposition by aligning your development with the city's master plan, sustainability goals, and economic benefits.
- Risk factors (flood, crime, political headwinds).

Example: Show how your auto condo project reduces illegal street parking, brings taxable value to the city, and promotes local luxury tourism.

3. POLITICAL & COMMUNITY INTELLIGENCE – INFLUENCE WITH PRECISION

Key Principle: Know who to call before the fire starts.

AI Execution:

- Stakeholder Mapping: AI identifies decision-makers, planning board members, and influential local players based on voting patterns, campaign contributions, and public commentary.
- Hyper-Personalized Outreach: Create tailored messaging that resonates with each stakeholder's publicly known priorities. Use AI to simulate their likely objections, and arm your team with preloaded rebuttals.

4. SPEED = LEVERAGE

Be first. Be faster. Be inevitable.

AI gives you:

- 10x faster site vetting
- 5x more accurate zoning risk profiles
- Real-time adjustment tools as the battle shifts

5. FINANCIAL MODELING & CAPITAL STACK ENGINEERING

AI builds and tests:

ASSET	TOOL	OUTCOME
Brand Video	Runway ML + ElevenLabs (voice)	Authority emotion, prestige
Website	Framer + AI copywriter (Jasper/ChatGPT)	Funnel + scarcity
Press Kit	ChatGPT + Canva	Investor credibility
Instagram Grid	Midjourney visuals + scheduling AI	Aspiration, attnetion
LinkedIn Articles	Chat GPT + competitor benchmarking	Thought leadership
Email Sequences	ActiveCampaign + AI copy	Nurture + scarcity conversion

- Pro forma scenarios
- IRR waterfalls
- Tokenized capital stack options
- Recession stress tests
- Optimal investor-return blends

Stop hoping your cap table works. Hope is not a strategy. The CoreBrain.ai runs 10,000+ capital structure simulations in seconds—stress-testing every equity move, dilution scenario, valuation trigger, and exit pathway. It calculates investor returns, control thresholds, and payout models across multiple deal timelines, then tells you exactly what works and why. No guesswork. No spreadsheet roulette. Just clarity and control.

Most founders hand away equity blindly, chasing money instead of structuring leverage. But real entrepreneurs—those building scalable, fundable companies—simulate every outcome before signing a term sheet. That's how you attract the right investors, protect your upside, and scale without sacrificing control.

This is not theory. It is financial architecture for long-term domination.

If you can't simulate it, you can't scale it. If you can't scale it, you're just gambling.

This is how you protect power. Make it your standard.

6. BRANDING & MARKETING STRATEGY

AI maps:
- Buyer personas with psychographic precision
- Content that resonates emotionally
- Behavioral ad targeting and funnel optimization

You're not marketing to the masses. You're whispering directly to the few who are predisposed to buy.

7. EXECUTION & PROJECT MANAGEMENT

AI tools forecast:
- Budget overruns
- Delay risks
- Resource allocation issues
- Supply chain disruptions
- Automate reporting, procurement reminders, and contractor coordination

Think of it as a 24/7 Chief Operating Officer that doesn't sleep, and will never forgets a task.

8. EXIT, LICENSING, & REPLICATION

AI models:
- Optimal time to exit
- Who is likely to acquire and why
- What markets you should franchise or license into next
- What value your brand/IP/data add to the asset

AI makes your first success repeatable, and your exit irresistible.

KEY TAKEAWAY:

Using AI in strategy and development doesn't replace vision; it amplifies it. Strategy and development is about building smarter, faster, and with less guesswork. When applied to specialized developments like auto condominiums, AI turns a boutique concept into a scalable, data-validated asset class.

KEY TAKEAWAY:

If you're not using AI in site selection and zoning, you're playing with one eye closed. It is about eliminating emotion, guessing, and politics; and replacing it with data-driven certainty.

CHAPTER 8

AI-POWERED PRECISION IN AUTO CONDO DEVELOPMENT

*"Raising capital isn't about selling.
It's about framing, status, and certainty." – JT Foxx*

"The most valuable investor is the one who aligns with product, timing, and trajectory. It's strategy—not volume—that fuels sustainable capital." – William Böll

In specialized real estate like auto condominiums, a hybrid between commercial warehousing and luxury man cave or she shed culture, success isn't about luck or instinct. It's about data, leverage, and speed of execution. In today's climate, guessing wrong means you're overbuilt and out of capital. Guessing right means you dominate a niche most investors don't even understand. Artificial Intelligence is not a buzzword—it is your arbitrage engine. Use it to eliminate speculation, enter only the right markets, and build only what sells.

STEP 1: IDENTIFY THE RIGHT MARKETS USING AI

Market Filter Framework (Enhanced with AI)

Before choosing where to build, you need to run every city or region through a 6-factor filter. Here's how to combine strategic criteria with AI tools:

1. AFFLUENT TOY OWNERS

Look for zip codes where household income is $150K+ and the percentage of residents with RVs, boats, classic cars, or luxury vehicles is 3x higher than the national average.

Use AI-enabled demographic databases like Placer.ai or predictive analytics through ERSI's Tapestry Segmentation.

2. LIMITED STORAGE INFRASTRUCTURE

Use AI to scan public records and MLS listings to locate underserved marketplaces with high demand and no premium storage supply.

Cross-reference zoning maps and industrial permits issued in the past 5 years. If you see massive growth without corresponding storage solutions, that's your gap.

3. REAL-TIME DEMAND SIGNALS

Use tools like Google Trends, Facebook Ads Library, and AI ad intelligence tools to see where consumers are already searching for RV, boat, and collector car storage.

AI can predict demand spikes 6–12 months out. Have data tell you where money is moving.

4. BUSINESS OWNERSHIP DENSITY

Garage condos aren't just for toys, as they attract business owners who want secure, zoned, write-off-ready workspace.

Target counties with high LLC registrations, SBA loans per capita, and service-based entrepreneurs.

5. DEVELOPMENT-FRIENDLY CITIES

Use AI to analyze zoning codes and permit turnaround times.

Shortlist cities with fast-track permitting and no resistance to light industrial use, which saves 6–12 months in holding costs.

6. HIGH-CAR-CULTURE REGIONS

AI tools can scrape and analyze event listings, club memberships, and traffic to car show sites.

Markets with car culture events every month = sticky emotional buyers.

Outcome: You now have a ranked list of the top 5–10 metro areas with optimal buying power, demand density, and low competitive friction.

STEP 2: USE AI TO PREDICT THE RIGHT UNIT COUNT

A mistake most developers make? They guess. They ask a broker. They "feel" out the market. That is not how empires are built.

Use AI to precision-size your development:

1. REVERSE ENGINEER ABSORPTION RATES

- Input local industrial sales data, days-on-market, and lease-up times into AI modeling tools (e.g., CoStar, Reonomy)
- Predict how fast units will sell based on historical velocity and micro-trend indicators (seasonality, local inventory, buyer sentiment)

2. MODEL PURCHASE PERSONAS

Use AI CRM tools to analyze sales from similar developments in other markets—profile who is buying (contractors, collectors, digital nomads).

Based on persona density in your market, AI estimates how many will convert.

3. SIMULATE PRICING SCENARIOS

Run AI simulations to test three variables: unit size, price per square foot, and fit-out customization options.

Based on price sensitivity analysis, you will know whether to build 24 premium units or 60 mid-market ones and which mix maximizes margin per square foot.

4. FACTOR IN SCARCITY ECONOMICS

JT always says, "Being sold out is marketing."

Don't build 100 units if the market will buy 60 fast and push scarcity premium pricing.

AI helps you identify the inflection point where demand drops off. Your sweet spot is 80% of peak demand.

STEP 3: BUILD TO THE EXIT

You're not building to own. You're building to flip, lease, or exit to a REIT or private equity group. AI helps optimize the development exit:

- Predict cash-on-cash returns under multiple sale timelines
- Run buyer targeting models to identify roll-up investors or private buyers with acquisition mandates in your niche
- Score each project against REIT acquisition criteria: lease rates, renewal trends, and regional portfolio fit

FINAL WORD: DON'T JUST BUILD. ENGINEER WITH PRECISION.

Auto condo development isn't about construction; it's about intelligent speculation. Your capital is at war with time. AI is the general—but you're still the decision-maker.

Use the data. Trust the model. Execute faster than your competition.

CHAPTER 9

THE JT FOXX AI SALES SYSTEM— FROM COLD LEADS TO CLOSED UNITS IN 21 DAYS

How to scale high-ticket real estate sales
without building a sales team

*"Execution is a discipline, not an idea.
And AI makes that discipline lethal." – JT Foxx*

*"Execution isn't about speed.
It's about eliminating rework, delay, and drag—
profit lives in that margin." – William Böll*

Developers think in timelines. Closers think in triggers. You don't sell faster by hiring more salespeople. You sell faster by controlling the psychology, positioning, and cadence of your buyer journey and then letting AI do the heavy lifting.

JT Foxx doesn't sell with pressure. He sells with positioning. This chapter gives you the entire system to convert cold leads into closed deals in 21 days or less using an AI-assisted sales framework without wasting time on tire-kickers, price shoppers, or people who "need to think about it."

1. REAL ESTATE SALES IS NOT ABOUT PERSUASION. IT IS ABOUT PRECISION.

Sales is a science. Every lead follows a behavior pattern. Every buyer responds to structured sequences. AI gives you what most closers never get, which is predictive insight and behavioral leverage.

You are not selling garages. You are closing identity-driven buyers who need to be triggered, not convinced.

2. THE JT FOXX SALES FRAMEWORK TEACHES PITCH, SELL, CLOSE (PSC) PLUS AI LAYER

Here's the framework with AI integration:

PHASE	GOAL	AI SUPPORT
Pitch	Frame the value	AI video, social content, brand funnel
Sell	Trigger desire	Email automations, retargeting, scarcity
Close	Remove friction	AI chatbots, scheduling, real-time objections

3. AI-POWERED BUYER ONBOARDING SEQUENCE (DAY 0–7)

From first contact to qualified conversation.

FUNNEL BUILD:
- Use Framer or ClickFunnels + ChatGPT to build a single offer page:

"A private automotive sanctuary for the elite—now accepting 12 qualified owners."

ENTRY POINTS:
- Instagram swipe-ups (visual branding → landing page)
- AI retargeting from Facebook and YouTube (luxury car enthusiasts)
- Email list: "Invitation-Only Founder Tier—5 Units Remaining"

ON-SITE:
- Embedded AI bot (ManyChat or Drift): qualifies in under 90 seconds

- Segment: Collector / Investor / Both

Buyer scores above 75% = instant call booking with Calendly + custom AI email reminder

It is not about chasing leads. It is about filtering them to the top.

4. EMOTIONAL TRIGGER CADENCE (DAY 8–14)

Buyers don't commit with logic. They tip over with emotional triggers. Use AI to automate 7 days of structured persuasion:

DAY	TRIGGER	DELIVERY
1	FOMO: "Units going fast"	SMS + image render
2	Social proof: Owner spotlight	Email + 30-second video
3	Authority: PR article drop	LinkedIn post + link
4	Lifestyle: Private event teaser	IG carousel + invite
5	Scarcity: Tier pricing rising	SMS w/timer
6	Personal note from the founder	AI-written email
7	Final call offer (upgrade bonus)	Email + AI video walkthrough

AI tools to use:
- ChatGPT for email writing
- Midjourney + Runway for visual assets
- Voice tools like ElevenLabs for custom follow-up messages

5. THE 21-DAY CLOSE CALL SCRIPT (JT STYLE)

"This isn't a garage. Be part of a legacy-level brand asset for

elite collectors. We're only moving forward with the right fit."
Why this works:

- Elevates exclusivity
- Positions the buyer as being selected
- Frames the conversation around identity, not inventory

Close Trigger Phrases:

- "We're looking for the right energy inside this club, not just capital."
- "You're early enough to lock Tier 1 pricing, but that ends this Friday."
- "The top three upgrades were just claimed. Want to see what is still available?"

Close with framed control, not force.

6. AI OBJECTION HANDLING + RE-ENGAGEMENT (DAY 15–21)

Most leads don't say no. They stall. AI helps revive without you chasing.

Common Objections & AI Responses:

OBJECTION	AI RESPONSE
"Need more time"	Time-sequenced scarcity email: "Only two founder units left"
"Still considering"	Personalized video: 30-second recap of the top emotional drivers
"Talk to spouse"	AI-scripted joing-decision email: "Why couples buy together"
Ghosting	"AI SMS: "We have reopened one Founder Unit for someone like you."

Automation Tools:

- HighLevel CRM for multi-channel nurture

- Jasper + ChatGPT for objection scripting
- Loom AI for follow-up video walk-throughs

7. POST-CLOSE AUTOMATION FOR REFERRALS & UPGRADES

After the deal is signed:

- Auto-send "Welcome Owner Kit" (AI-personalized video + PDF)
- Trigger upgrade upsell: "Interior concierge package available for 3 more buyers"
- Request testimonial via video AI (drop in name, add music and branding)

Referral Prompt (AI-enhanced):

"Do you know one person who should be on this journey with you? We will fast-track their application with your referral."

Incentivize with:

- Free design upgrade
- Event invite
- Lifetime membership bump

BOTTOM LINE:

You don't need a massive sales team. You need a precision-engineered system that triggers emotion, filters fast, and closes with status.

With AI, you can scale the JT Foxx method without scaling cost, complexity, or compromise.

CHAPTER 10

YOUR MILLION-DOLLAR PRECISION TOOLS

"You're not selling square footage. You're selling identity, status, and emotion." – JT Foxx

"We don't sell real estate. We sell certainty, brand, and identity—AI just ensures we deliver the message with precision." – William Böll

In real estate development, every mistake has a price tag, which is usually written in seven figures. In specialized niches like auto condominiums, your margin for error is zero. If you guess, you bleed. If you assume, you lose.

That's why building information modeling (BIM) and bill of materials (BOM) aren't just construction tools; they are weapons of strategic execution. Used correctly, they compress timelines, control costs, increase profit, and unlock investor confidence. Ignore them, and you're flying blind into a war zone.

This chapter is your blueprint to using BIM and BOM, not like an architect, but like a real estate investor building for margin, velocity, and an engineered exit.

1. BIM: YOUR DIGITAL TWIN FOR STRATEGIC EXECUTION

What It Is: A Building Information Model is a dynamic, 3D digital replica of your development. But unlike static blueprints, BIM holds live, data-rich objects, not just geometry.

Why We Use It: Because margins live in the details that most developers overlook. With BIM, you can forecast everything before you pour a single yard of concrete.

HERE'S WHAT BIM GIVES YOU IN AUTO CONDO DEVELOPMENT:

Clash Detection Before Construction:

- No surprises in HVAC versus electrical. No costly rework. BIM flags every conflict before it happens.

Pre-visualized Lifestyle Upsells:

- You're not selling a garage; you're selling an experience. Monaco Editions. Aspen Fit-Outs. Dubai Detailing Bays. BIM lets your buyer see it before they buy it.

Phased Monetization Modeling:

- Do you want to release in three phases and raise pricing for each tier? BIM shows you how each phase impacts the rest of the site in real-time.

Contractor Bidding Leverage:

- With BIM, contractors bid on precisely what is visualized. You eliminate ambiguity, which kills change orders and inflates your costs.

Tokenization Clarity:

- If you're raising capital through tokenization or structured syndication, BIM is your credibility asset. Investors see what they're buying into, not abstract paper plans.

JT Foxx Rule: "If you can't simulate it, you can't scale it. If you can't scale it, you're just gambling."

2. BOM: YOUR LINE-BY-LINE PROFIT PRESERVER

What It Is:

- The Bill of Materials is your comprehensive inventory of every physical component in your build from rebar to finish screws. It's the DNA of your construction costs.

Why Most Developers Ignore It:

- Because it's "too detailed." When they are $2M over budget, staring at a 6-month delay and trying to figure out where it went wrong, it becomes less detailed and critically important.

Why We Treat It Like a Financial Statement:

- Because the BOM is your cost control engine and material timeline optimizer. It tells you what to buy, when to buy it, and how to hedge against volatility.

How BOM protects your auto condo project:

- Real-Time Cost Tracking. Your BOM integrates with procurement and BIM. One shift in global steel pricing and you can instantly model its impact on your entire shell budget.
- Lock-In Pricing. Early BOM lets you pre-buy 30–40% of materials most exposed to inflation risk, which protects your margin before the first trench is dug.
- Vendor Negotiation Weapon. Bring a BOM to your supplier, and you have leverage. You're not a builder; you are a logistics machine with a purchase strategy.
- Unit Cost per Square Foot (sf^2) Optimization. You want to build at $263/$sf^2$ and sell at $463/$sf^2$. BOM is how you build that spread. It shows you exactly what is driving up the cost per square foot and where to substitute or eliminate.
- Scalable Fit-Outs. When your lifestyle upgrades include lift systems, mezzanine lounges, or electric vehicle (EV) charging, the BOM lets you modularize each upsell into a repeatable, trackable stock keeping unit (SKU).

3. COMBINED POWER: BIM + BOM = CAPITAL TRUST

This is your real leverage: When investors, lenders, or toke-nized capital partners review your deck, they are not just looking at projected profits. They are asking:

- Can this team execute on time and on budget?
- Are these numbers rooted in assumptions or real data?
- What is their control mechanism when costs spike or supply chains are delayed?

BIM + BOM is your answer. It proves you are not just a dreamer. You are a builder with systems, contingencies, and execution mastery.

Developer Action Plan

4. Invest in a BIM Partner. Don't let your general con-tractor (GC) or architect control the BIM. You control the file, the data, and the standards. It is a development asset, not a drawing.

5. Build a Modular BOM Library. Every future project should reuse 60–80% of the same BOM struc-ture. You're not just developing, you are pro-ductizing your builds.

6. Integrate BIM/BOM with Pro Forma. Every line in your model should map back to revenue or cost centers in your financial stack, which is how you create investor confidence.

FINAL WORD

Most developers still build like it's the 90s. They react to prob-lems on-site. They hope suppliers come through. They pray that framing costs stay flat.

That's not business; that's chaos in slow motion.

Use BIM to plan every phase. Use BOM to protect every dollar. Let your competitors overbuild and overspend.

You? You will build with precision, scale with confi-dence, and exit with dominance.

CHAPTER 11

TOOLS YOU MUST USE (BEYOND BIM & BOM)

"The smartest exit is the one you planned before you even raised a dollar." – JT Foxx

"We build with the end in mind. The right exit isn't just about valuation—it's about velocity, reputation, and leverage." – William Böll

1. REAL-TIME COST MANAGEMENT SOFTWARE

Tool Examples: Procore, Buildertrend, CoConstruct

Why You Need It:

- Tracks every cost in real time, mapped against your BOM and timeline
- Red-flag notifications when you're trending over budget
- Integrates invoices, bids, and vendor change orders with no surprises and no leaks

JT Principle: "If you don't know your daily burn rate, you are already behind."

2. AI-BASED DEMAND FORECASTING & SITE ANALYTICS

Tool Examples: JT Foxx CoreBrain.ai, Placer.ai, Reonomy, Local Logic

Why You Need It:

- Pinpoints which submarkets have high-income hobbyists, limited commercial zoning, and rising traffic in toy storage demand

- Combines foot traffic, demographics, and psychographic data for custom buyer personas

JT Principle: "Data kills emotion. Let the algorithm show you where the money is moving."

3. PRE-SALES CRM WITH MARKETING AUTOMATION

Tool Examples: HubSpot, Salesforce, Zoho CRM + Mailchimp or ActiveCampaign

Why You Need It:

- Tracks every prospect, conversion point, and upsell opportunity from day one
- Automates follow-up sequences based on behavior (clicks, opens, inactivity)
- It is critical when building buyer urgency around scarcity tiers

JT Principle: "Don't market to strangers. Market to a database that's been built for profit."

4. DIGITAL TWIN & VIRTUAL WALKTHROUGH TECH

Tool Examples: Matterport, Unreal Engine, Twinmotion

Why You Need It:

- Sell the lifestyle before concrete is poured
- Let investors and buyers tour units digitally with customized fit-outs (Monaco, Aspen, etc.)
- Increases pre-sale pricing and compresses the sales cycle

JT Principle: "If they can see it, they'll pay more for it. If they can feel it, they'll close faster."

5. CONSTRUCTION SCHEDULING & WORKFORCE MANAGEMENT TOOLS

Tool Examples: Smartsheet, Primavera P6, Fieldwire

Why You Need It:

- Visual Gantt timelines that sync with contractors, vendors, and city inspections

- Tracks workforce allocation, subcontractor milestones, and material drop schedules
- Reduces holding costs by weeks or months if managed aggressively

JT Principle: "Time is the only cost you can never get back. Win the schedule, win the project."

6. INVESTOR REPORTING & TOKENIZATION PLATFORMS

Tool Examples: Vertalo, DigiShares, Securitize

Why You Need It:

- If you're tokenizing equity or structured cash flows, you need bulletproof compliance and smart contract deployment
- Ensures Class A investors get real-time updates, quarterly yield, and visibility into governance decisions
- Increases liquidity, widens capital pool, and removes bottlenecks from traditional capital raising

JT Principle: "Your ability to attract capital is proportional to how well you manage trust and transparency."

7. VENDOR MANAGEMENT & PROCUREMENT PLATFORMS

Tool Examples: Kojo (fka Agora), GEP SMART, SAP Ariba

Why You Need It:

- Bulk material pricing, competitive bids, lead-time tracking
- Tracks supplier performance and shipping risk exposure
- Avoids the 5–10% hidden cost creep on every project.

JT Principle: "Every dollar you don't track becomes someone else's margin."

8. SMART ENTITLEMENT & PERMIT TRACKING TOOLS

Tool Examples: Symbium, PermitFlow

Why You Need It:

- Helps speed up entitlements by visualizing zoning compliance and application timelines
- Identifies hidden land use risks and delays before acquisition
- Tracks every approval stage, inspector appointment, and city comment

JT Principle: "Permits don't delay your project. Poor preparation does."

FINAL WORD: THE STACK BUILDS THE SPREAD

Every elite real estate developer has two things:

- A system that protects their margin
- An integrated toolset that eliminates inefficiency

If you're going to compete with institutional capital, hedge funds, and vertically integrated GCs, your advantage is speed, clarity, and data. That's what this stack gives you.

BIM and BOM are just the foundation. Everything else turns it into a profit machine.

CHAPTER 12

COST ENGINEERING WITH AI TOOLS

How to reduce waste,
optimize every dollar, and
scale profitably using artificial intelligence

*"You either build with integrity or collapse
under exposure. There's no middle ground." – JT Foxx*

*"Profit without principle is short-lived.
We build systems that win in the market and
stand the test of scrutiny." – William Böll*

AI USE CASES:	GOAL
AI-Powered Feasibility	Predict land value and ROI using zoning + collector data
Generative Layout Design	Auto-generate luxury unit layouts with optimal cost per sf^2
Trade Bid Optimization	AI reviews trade bids for pricing, risk, and performance
Value Engineering	Model how finish/material swaps impact perceived luxury
Timeline Compression	AI sequences Gantt charts to reduce build time
Supply Chain	AI Forecast material lead times, optimize procurement

Cost engineering with AI tools is how elite developers slash waste, compress timelines, and increase margins—exactly how we approach real estate like auto condo developments. It is not about saving pennies; it's about engineering profitability from day one by turning uncertainty into calculated control. This approach is core to the Velocity Performance Alliance system of delivering high-margin, high-status auto condominium assets.

1. AI-DRIVEN BUDGET MODELING — PRECISION OVER PROJECTIONS

Key Principle: Guessing is gambling. Engineering is wealth.

Use AI platforms to:

Analyze thousands of similar builds to generate hyper-accurate cost projections for site work, materials, and labor

Simulate budget scenarios based on land size, location, material volatility, and contractor history

Forecast escalation risks (steel, concrete, HVAC) and lock pricing windows with suppliers

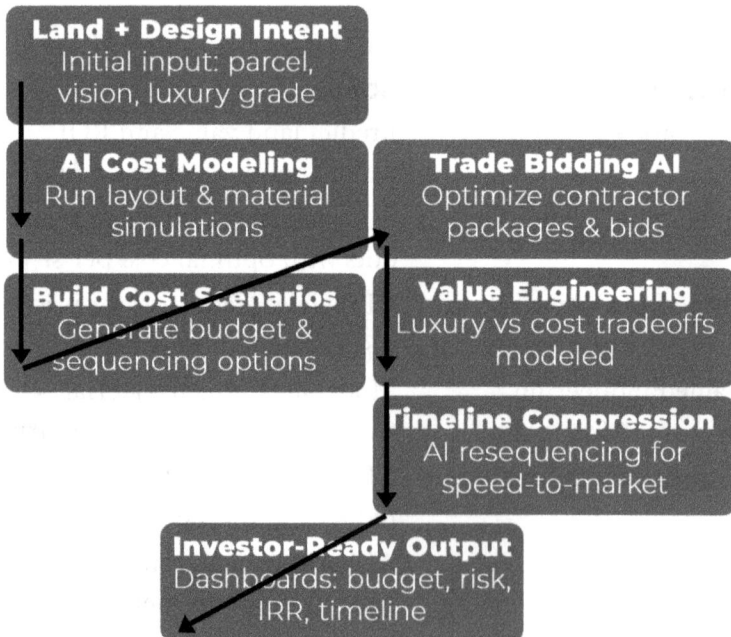

Land + Design Intent
Initial input: parcel, vision, luxury grade

AI Cost Modeling
Run layout & material simulations

Trade Bidding AI
Optimize contractor packages & bids

Build Cost Scenarios
Generate budget & sequencing options

Value Engineering
Luxury vs cost tradeoffs modeled

Timeline Compression
AI resequencing for speed-to-market

Investor-Ready Output
Dashboards: budget, risk, IRR, timeline

Predicts overruns before they happen

Optimizes value per dollar spent

Enhances decision-making with predictive data models

Result: You prevent budget creep and negotiate harder, earlier.

2. VALUE OPTIMIZATION — NOT COST CUTTING, STRATEGIC SUBSTITUTION

AI Execution:

AI tools recommend value-equivalent substitutes for finishes, insulation, and structural components that cut cost without reducing perceived or real value

Real-time side-by-side cost versus quality impact visuals

Supply chain integration to see what's locally available or bulk-discounted regionally

Key Tactic: Swap luxury items that look expensive but cost 30% less, which maintains margin while elevating perceived value.

3. CONSTRUCTION TIMELINE ENGINEERING

AI analyzes historical cost data, real-time supplier rates, and location-based pricing to create hyper-accurate budgets within seconds.

Key Principle: Time is money. Delays are cancer.

AI can:

- Sequence project schedules for maximum efficiency (foundation to finish)
- Predict delays based on historic weather, supplier reliability, and inspection timing
- Dynamic pricing from millions of projects
- Labor and materials forecasts by region
- Adaptive estimates based on design changes
- Flag contractor underperformance or bottlenecks before they escalate

Use Case: If drywall delays show up two weeks early in the AI model, you renegotiate or reassign instantly before the GC causes a domino delay.

4. REAL-TIME COST TRACKING + ANOMALY DETECTION

AI monitors your live budget feed:

- Flags billing anomalies, overcharges, or non-contracted change orders
- Tracks contractor efficiency versus industry benchmarks
- Warns when unit economics deviate from your master pro forma

Our Strategy: This eliminates surprises. Your investors see disciplined execution. That raises your credibility and future capital velocity.

5. DESIGN-TO-COST ENGINEERING (DTC)

Before permits are filed, AI reverse-engineers design features to fit a hard cost target:

- Want a $210 per sf^2 build cost? AI will design backward to hit it
- Integrates layout, finish level, and MEP constraints instantly
- Turns speculative planning into tactical execution

6. DESIGN-TO-COST AUTOMATION

AI-powered design tools (e.g., generative design, BIM with ML overlays) can:

- Redesign building elements to reduce structural waste
- Recommend cheaper or more sustainable materials
- Model cost-saving layout changes in real time

Key Example: Can we save $600K if we rotate the structure 90°? AI answers before you finish the sentence.

7. CONSTRUCTION SCHEDULE OPTIMIZATION

AI tools like Artificial Linguistic Internet Computer Entity (ALICE) or Buildots simulate different construction sequences to:

- Reduce labor downtime
- Shorten project timelines
- Sequence deliveries for minimal delays

Key Principle: Shave weeks off your build without compromising quality.

8. PROCUREMENT INTELLIGENCE

AI predicts supply chain delays, price spikes, or alternative vendors by:

- Scraping procurement databases
- Monitoring geopolitical events and logistics routes
- Recommending just-in-time delivery strategies

Key Principle: Never get caught off guard by a material shortage again.

9. CHANGE ORDER PREDICTION & PREVENTION

AI can flag areas of your plan most likely to trigger change orders before they happen.

- Tracks plan/design inconsistencies
- Compares with databases of similar past mistakes
- Warns teams in advance

Key Principle: Prevent the number one profit killer in development: unexpected changes.

9. CASH FLOW & BUDGET HEALTH DASHBOARDS

AI-integrated dashboards offer:

- Real-time burn rate tracking
- Variance alerts
- Forecasted runways and liquidity cliffs
- Drill-down views for cost category audits

Key Principle: Turn financial chaos into a single, visual source of truth.

10. RESULTS WHEN AI IS USED IN COST ENGINEER- ING:

- 15–25% reduction in avoidable costs
- 2–6 months shaved from total build cycle
- Higher investor confidence from precision budgeting
- Better risk-adjusted returns for LPs

11. ESPECIALLY POWERFUL FOR SPECIALIZED DE- VELOPMENTS:

For auto condominiums, branded lifestyle assets, or medical retreats, where fit-and-finish matters and design is unique, AI allows you to scale luxury without overbuilding.

Key Takeaway:

Cost engineering isn't about being cheap. It's about controlling the outcome before a dollar is spent.

CHAPTER 13

EXIT ENGINEERING— LICENSING, REPLICATION, & GLOBAL SCALING

How to turn one development into a global asset without starting from scratch

"Blueprints are nothing without execution. Build it, scale it, repeat." – JT Foxx

"A profitable real estate product is engineered. It doesn't emerge by accident. That's our method. That's our advantage." – William Böll

Anyone can build a building. Elite operators build platforms.

While most developers obsess over units sold, JT Foxx focuses on scalable exits—replicable IP, global license models, and engineered brand equity that compounds in value after the first build.

You don't get wealthy from the project. You get wealthy from what the project enables.

This chapter unpacks the successful strategy to turn your specialized development into a repeatable, investable, and exit-ready machine, using licensing, global partner expansion, and brand equity monetization—powered by AI.

1. BUILD ONCE. SELL FOREVER.

The average developer sells a project and moves on. We build it once and sell it dozens of times through:

- Brand licensing
- Franchise-style expansion
- JV models with operators in other markets
- Packaging IP + systems as a business asset

The goal: Exit from operations. Keep the upside.

2. OUR EXIT MODEL (L.R.E. FRAMEWORK)

PHASE	DESCRIPTION	OUTCOME
L	License the brand and model	Sell use of the name + system
R	Replicate in new geographies	Launch in parallel with local capital
E	Exit via equity stake or packaged roll-up	Monetize brand, not just units

3. PACKAGE YOUR IP AS AN ASSET

Buyers don't just want buildings. They want blueprints.

Here's what you must codify:

- Deal structure and legal templates
- AI-driven marketing system (funnels, brand, social)
- Construction timeline + vendor lists
- Investor deck templates
- Sales system (scripts, tools, objection handling)

Use Notion + Airtable + Google Drive as your backend repository. AI can auto-generate standard operating procedures and investor pitch templates.

You have turned your "project" into a plug-and-play business model, which is precisely what institutional buyers or JV partners want.

4. BRAND LICENSING: GLOBALIZE WITHOUT STARTING OVER

Instead of building again, license your brand.

How it works:
- You own the IP + brand identity
- They (local partner) bring land + capital
- You supply the system (via AI-powered digital back-end) and they pay:
  - Upfront license fee
  - % of gross sales
  - Optional consulting retainer

You become the "Ferrari" of auto condominiums. They build the car; you supply the badge.

AI automates:
- Training portals
- Weekly reporting templates
- Launch playbooks for each new city

5. THE 4 LEVELS OF EXPANSION STRATEGY

LEVEL	STRATEGY	CAPITAL NEEDED	YOUR ROLE
1	Internal replication (same team, new site)	High	Full ops
2	JV partner model (land + capital)	Medium	Oversight
3	Licensing (branding + systems only)	Low	Advisory
4	Global roll-up (consolidated portfolio sale)	Varies	Exit

Move from Level 1 to 4 as you build equity in your brand, not just your buildings.

6. INSTITUTIONAL BUYERS WANT BRAND ASSETS

What private equity and REITs want:
- Scalable asset classes
- Repeatable operations
- Strong buyer loyalty
- Data and behavioral patterns
- Exit predictability

Your job: Package the business, not just the real estate.

AI supports this by generating:
- Automated investor reports
- Branded deal decks
- Data dashboards (sales cycle time, marketing ROI, IRR by project phase)

Do not position yourself as a developer, but as a portfolio architect.

7. GLOBAL PARTNER TARGETING — POWERED BY AI

AI helps you identify ideal license partners by scanning:
- Existing developers in secondary luxury markets
- High-net-worth individuals with a passion for autos/ lifestyle investing
- Franchise operators looking for niche brand expansions

Tools:
- LinkedIn Sales Navigator + ChatGPT prompt stacks
- Apollo.io for email lists + verified contact data
- SparkToro for affinity profiling by brand or interest

Send AI-generated outreach sequences that position your opportunity as exclusive, prestige-backed, and turnkey.

8. CREATE SCARCITY AT THE LICENSING LEVEL

Don't offer licenses to everyone. Frame them like country-level master franchises.

"We're offering ONE license per region. You'll control the territory, the IP, and the rollout schedule."

Include:
- Term-based exclusivity
- Minimum build quotas
- Brand compliance standards

This increases perceived value, licensing fees, and long-term brand control.

9. STACK YOUR EXIT OPTIONS (REAL PLAYBOOK)

Here's what a real exit stack looks like:

PHASE 1: PERSONAL CAPITAL RAISE
→ Build first flagship project

PHASE 2: USE AI + BRANDING TO DRIVE FULL PRE-SELL
→ Cash-on-cash returns locked before completion

PHASE 3: LICENSE BRAND TO 3–5 REGIONS
→ Collect 5–7 figures per license + % of sales

PHASE 4: PACKAGE ALL REGIONS INTO A ROLL-UP
→ Pitch to private equity as a scalable, niche vertical

PHASE 5: FULL EXIT OR RETAIN BOARD POSITION + MINORITY EQUITY
You're not exiting a building. You're exiting a system—branded, proven, and globally positioned.

BOTTOM LINE:
- The goal isn't to sell a project. It's to sell the machine that creates projects. That's where exponential wealth is built. That's where scale happens without stress.

AI gives you:
- The systems. Brand equity gives you the leverage. Execution gives you the exit.

CHAPTER 14

BUILDING THE AI DEVELOPER DASHBOARD— REAL-TIME CONTROL, AUTOMATION, AND DATA-DRIVEN COMMAND

How to run a lean development empire
without losing control or oversight

Real developers track construction. Elite developers command the business in real time.

You don't build scalable wealth by managing spreadsheets and chasing vendors. You build it by engineering a centralized command system that gives you live data, strategic foresight, and the power to control capital, marketing, construction, and investor relations—all from one place.

This chapter outlines the Developer Dashboard—a fully integrated, automation-powered cockpit for specialized real estate execution. Built lean, scaled smart, and controlled from the top down.

1. WHY EVERY DEVELOPER NEEDS A CENTRAL AI DASHBOARD

If you're managing deals across:

Capital stacks

Vendors

Marketing channels

Sales pipelines
Zoning boards
Investor relations
...and you're doing it in email, Slack, Excel, and WhatsApp...
You're bleeding time, control, and money.
JT's Rule: "If I can't see it in 90 seconds, I don't own it."
You need a single-source command center.

2. THE AI DEVELOPER DASHBOARD — CORE STACK OVERVIEW

Platform Base:

Notion or Airtable (for custom architecture)

Integrated with:

Zapier or Make (for automation)
OpenAI/generative (for decision support + auto-analysis)
Google Sheets + Data Studio (for real-time visualization)
This dashboard tracks:

FUNCTION	WHAT IT CONTROLS
Capital	Investor commitments, funding tranches, and ROI forecasts
Sales	Pipeline, close rate, top performers, objections
Marketing	Funnel traffic, lead sources, conversion %
Construction	Milestone status, vendor performance, burn rate
Zoning	Status, risks, approval timeline
AI Automations	Triggers, alerts, system performance

3. CAPITAL COMMAND CENTER

Purpose: Track and optimize investor engagement, funding velocity, and ROI commitments.

AI CAPABILITIES:

Auto-update investor dashboards

Send personalized updates via email + text

Highlight capital gaps + trigger raise sequences

AI suggests raising priorities based on IRR forecasts and burn rates.

4. MARKETING + SALES FUNNEL TRACKER

Purpose: View lead flow, friction points, and emotional triggers in real time.

WHAT TO TRACK:

Entry point by channel (ads, referrals, PR)

Conversion by segment (collector, investor, or both)

Lead score + close probability

Follow-up sequences fired (email/SMS)

Tier selection breakdown

AI USE:

Suggest optimized offers

Trigger time-sensitive scarcity campaigns

Forecast sellout velocity

5. CONSTRUCTION TRACKER (POWERED BY AI & OPENSPACE)

Purpose: Own the build process without micromanaging.

LIVE DASHBOARD INCLUDES:

Phase completion %

Drone/sensor visual feed (OpenSpace AI integration)

Labor issues flagged

Delays projected

CapEx versus plan variance

AI ALERTS:

"Phase 2 delayed by four days. Cost variance at 6%. Recommend vendor audit."

You no longer hope things stay on track. You get notified when they don't.

6. ZONING & LEGAL WATCHLIST

Purpose: Track all permits, rezoning, and political variables.

- Current zoning status
- Open hearings or council votes
- Risk flags (school proximity, public resistance)
- Lobbying activity + influencer map

AI Utility:

- Scrape municipal websites
- Auto-compile council agendas
- Draft position letters and pitch decks for political influence

7. AI AUTOMATION CONTROL CENTER

Where your real leverage lives.

Here's what to control with AI:

AUTOMATION	TRIGGER	OUTCOME
Investor Update	Capital milestone hit	Send video + update PDF
Lead Nurture	Form filled	Email + text + social retargeting
Close Reminder	3 days post-pitch	SMS + scarcity offer
Sales Call Recap	Call ends	AI generates summary + tasks
Construction Delay	3-day deviation	Alert + alternate vendor suggestion

All of this runs 24/7, with you watching from the top.

8. EXECUTIVE DASHBOARD SNAPSHOT (WHAT JT SEES)

Your personal home screen shows:

METRIC	EXAMPLE
Total Committed Captal	$8.7 M
Units Sold	17/24 (Founder Tier sold out)
Investor Net Promoter Score (NPS)	9.4
Construction Delay Index	Green (2-day buffer)
Marketing ROI	6.2x (Lead Source: Instagram)
AI Automation Status	94% successful last 7 days

If you can't get that in under two minutes, you don't have control.

9. AI AGENTS FOR DEVELOPER TASKS

Launch GPT-powered agents to handle routine tasks:
- Investor Agent: Answers LP questions 24/7 via chat
- Permit Tracker Agent: Updates zoning status weekly
- Content Agent: Posts branded social media content daily
- Vendor Audit Agent: Flags cost overruns or labor delays

You don't need a massive team. You need smart systems and sharp oversight.

BOTTOM LINE:

Build once. Automate aggressively. Scale without chaos.

The AI Developer Dashboard gives you total visibility, fast decisions, and the power to operate like a billion-dollar shop even with a lean team.

CHAPTER 15

AI FOR CAPITAL RAISING & SYNDICATION

AI for Capital Raising & Syndication isn't a gimmick; it is a power move to attract higher quality investors, close faster, and scale globally with precision. Raising capital isn't about spreadsheets. It's about confidence transference. AI gives you something most developers never have: certainty.

AI-backed modeling crushes objections before they're spoken.

You don't "ask" investors for money. You show them the data that makes their decision automatic. Run AI-powered heat maps to identify liquidity zones—layer pitch, sell, close (PSC) psychology in every presentation. Feed AI prompts with real objections from investor calls, and train rebuttals at scale. Investors don't buy into the building; they buy into you as the operator who knows how to weaponize information. AI is your edge. Use it to close before the shovel hits dirt.

Using AI in luxury auto condominium projects like Velocity Prestigious Auto Residences offers a significant strategic advantage. Below is a structured breakdown of how AI can be deployed to raise capital faster, target the right investors, and syndicate effectively.

1. AI-ENHANCED INVESTOR TARGETING

Key Principle: Right capital, right time, right structure.

Use AI tools to:

- Scrape and analyze public filings, real estate forums, social profiles, and private investor databases
- Identify accredited investors who've backed similar

asset classes (luxury storage, niche real estate, lifestyle plays)

- Build custom avatars of investors based on deal size, exit preference, and brand alignment
- Use machine learning and natural language processing (NLP) to identify and profile ideal investors:
- Investor Matching Algorithms: Match HNWIs, family offices, or private equity firms with your project based on past deal preferences, geography, asset class, and risk profile
- LinkedIn & CRM Mining: Scrape and score investor leads using AI-powered tools like Clay, Apollo, or Affinity
- Predictive Warmth Scores: AI can rank prospects by likelihood to engage or convert based on behavioral signals and prior interest patterns

Key Takeaway: You're pitching pre-qualified prospects who are already looking for a deal like yours.

2. CUSTOM PITCH DECK ENGINEERING — BUILT FOR CONVERSION

AI Execution:

- AI systems (like JT Foxx's CoreBrain.ai business stack) can generate pitch decks aligned to your investor profile (risk-tolerant, yield-focused, lifestyle-aligned, etc.)
- It tailors language, visuals, and emphasis to match investor psychology: IRR for funds, emotion for HNWI, comps for family offices
- It tests variations of deal positioning (e.g., "trophy asset" versus "cash flow play") based on conversion probability

Key Strategy: Don't tell your story. Engineer the one they want to hear.

3. AUTOMATED FOLLOW-UP & INVESTOR NURTURE

Key Principle: Speed + consistency = trust.

AI lets you:

- Automate follow-up emails, milestone updates, and value-add reports personalized for each investor
- Track engagement and interest signals (who opened the deck, which slide(s) they rewatched)
- Use chatbots to answer FAQs 24/7 and book calls when investor interest is highest

Outcome: You become the most professional, consistent, and responsive dealmaker in their inbox.

4. REAL-TIME DATA ROOM INTELLIGENCE

Once investors are in, AI:

- Flags questions asked by other LPs to preemptively create Q&A content
- Tracks who is in the data room and which material engaged them
- Suggests when to push closing calls based on behavior (not gut feeling)

Key Strategic Play: Shorten capital raise cycles by removing all perceived friction in due diligence.

5. AI FOR LEGAL STRUCTURING & SCENARIO TESTING

Leverage AI to assemble capital stacks and manage multi-party investment rounds:

- AI-Optimized Cap Table Modeling: Adjust equity splits, waterfalls, and returns to simulate multiple LP/GP scenarios
- Tokenization Engines (optional): Use AI-compatible blockchain tools to structure fractional shares or digital securities for easier syndication
- Automated Compliance Checks: Validate accreditation, regional limitations, and Know Your Customer (KYC) via AI-integrated legal tech platforms

AI platforms like Clause or LegalMation:

- Generate syndication-ready term sheets and operating agreements aligned to SEC guidelines
- Stress-test different waterfall models for LP/GP splits, preference returns, and promote triggers
- Simulate best-case/worst-case investor returns under different hold and exit scenarios
- Enter investor meetings with certainty, not speculation

6. AUTOMATED INVESTOR MATERIALS

AI cuts time from weeks to minutes:

- Pitch Deck Generators: Use GPT-powered design tools to rapidly customize decks for each investor type
- Custom Investor Portals: Use AI to auto-generate personalized dashboards with investment summaries, ROI projections, and deal documents
- Natural language Processing (NLP)-Powered Q&A Bots: Embed a bot in your data room that answers investor questions 24/7 based on your offering docs

7. BEHAVIORAL RETARGETING & CONVERSION

AI ensures you don't lose leads:

- Precision Retargeting: Serve custom ads or email follow-ups to investors who viewed your deck but didn't engage
- Conversion Analytics: AI tools like Mutiny, Hotjar + GPT-4o can interpret behavior and recommend changes to increase deal signups
- Lead Scoring Algorithms: Real-time scoring of investors to surface top-priority follow-ups based on email opens, call engagement, and portal activity

8. NARRATIVE ENGINEERING

Craft the perfect story using AI-trained persuasion models:

- Emotional Anchoring: Fine-tune your story to match the emotional triggers of each investor profile (status, legacy, exclusivity, or cashflow)
- Language Optimization: AI can rewrite your executive summary or pitch email in tones like "ultra-luxury," "family office-caliber," or "tech-forward cash-on-cash yield"
- Adaptive Ask: AI suggests optimal timing and ask structure (equity versus note versus convertibles) based on data and investor archetypes

9. AI-AUGMENTED SYNDICATION CRM

Build a real-time intelligence layer over your investor CRM:

- Deal Flow Dashboards: Track where capital is stuck and who needs nudging
- Next-Best Action Engines: AI alerts you to text, email, or call at the optimal moment with the right message
- Investor Sentiment Analysis: NLP detects hesitation or eagerness in investor emails and flags follow-up urgency

Key Takeaways:

- Raise capital 2–3x faster
- Cut investor material prep by 90%
- Syndicate smarter with higher transparency and trust
- Increase LP confidence with tech-driven reporting
- You don't just raise money. You engineer irresistible confidence. AI lets you raise smarter, close faster, and scale capital like a machine.

CHAPTER 16

AI-AUGMENTED SALES USING PSC

AI-augmented sales using pitch, sell, close (PSC) transforms your sales process from instinct-driven to precision-engineered

As described in JT Foxx's book *Business Is War: If You Want To Win, Learn From Failures, Not Success*, the PSC method already dominates because it's structured, persuasive, and duplicable. Now, layered with AI, it becomes unbeatable.

Here's how we execute AI-integrated PSC to close more deals, faster, and at higher margins:

1. PITCH: AI-OPTIMIZED HOOK & PERSONALIZATION

Key Principle: The first 7 seconds determine if they'll buy or block you.

AI Execution:

- Use AI tools to analyze the prospect's digital footprint (website behavior, social media, past purchases) and auto-generate a personalized opening line or value hook
- Run your pitch scripts through AI models to test variations for different buyer personas (emotional, analytical, skeptical, dominant)
- Voice and video AI can simulate tonality and delivery for maximum impact during live pitch training

Result: Deliver the right pitch to the right person with the right words every time.

2. SELL: DATA-DRIVEN PERSUASION THAT BYPASSES OBJECTIONS

JT Principle: Selling is not talking. It's triggering decisions.

AI Execution:

- AI recommends persuasion strategies based on prospect personality traits based on dominance, influence, steadiness, and conscientiousness (DISC) or Big Five model derived from language analysis.
- Based on CRM data, AI predicts objections before they surface—and delivers real-time rebuttal prompts for your sales representative or team.
- Sales content (case studies, videos, ROI calculators) is auto-sent based on where the prospect is in the buying journey—not randomly blasted.

Strategic Move: Show the client who they'll become with your offer, not just what they'll buy.

3. CLOSE: PREDICTIVE CLOSING INTELLIGENCE & FOLLOW-UP PRECISION

Key Principle: If you don't own the close, you lose the margin.

AI Execution:

- AI models (based on historical close rates, timing, and sentiment) predict close probability and suggest the best timing/method to re-engage
- Auto-generate custom closing offers (bonuses, urgency triggers, fast action rewards) based on buyer behavior
- Predictive scripts adjust in real-time based on buyer hesitation signals (hesitation = change language, not just push harder)
- Follow-up becomes a science:
  - AI delivers 1-to-1 follow-up messages tied to emotional drivers, objections, and priorities
  - Sequences are no longer guesswork; they are engineered closers

4. COACHING THE SALES TEAM (OR YOURSELF) WITH AI FEEDBACK LOOPS

Key Strategy: Refine your pitch like an athlete refines their swing

- AI tools (e.g., Gong, Chorus) analyze tone, pace, objection handling, and word choice across thousands of sales calls
- You receive weekly performance reports and AI-driven recommendations for each representative
- Auto-generated improvement tasks help representatives improve PSC execution faster

5. PSC + AI = PREDICTABLE, SCALABLE SALES

What this means in JT Foxx's PSC model:

- PITCH becomes customized instantly at scale
- SELL becomes a data-driven persuasion engine
- CLOSE becomes a predictive strike, not a hopeful guess

You're now closing 8 out of 10 without selling, just as JT teaches, but with a machine behind you doing the heavy lifting.

KEY TAKEAWAY:

- AI doesn't replace PSC. It amplifies it to levels human-only sales teams can't reach.
- "This is how JT Foxx would run a sales organization today: AI-enhanced, authority-driven, and closing with certainty."
- "Master the science. Control the psychology. Let the machine accelerate your success."

CHAPTER 17

THE INVESTOR MACHINE— HOW TO BUILD A PERPETUAL CAPITAL ENGINE POWERED BY AI

Raise faster. Retain control.
Repeat predictably.

Capital doesn't flow to the best projects.
It flows to the best-framed opportunities with systemized communication and predictable ROI logic.

Most developers chase investors.

Successful developers engineer an environment where investors chase the deal, and then keep coming back.

This chapter shows you how to build an AI-powered capital engine that raises money on autopilot, impresses investors with real-time reporting, and structures deals that scale you, not replace you.

1. CAPITAL IS NOT THE PROBLEM. CONFIDENCE IS.

Every market has dry powder. The issue is perceived risk and a lack of sophistication from the developer.

Your job isn't to ask for money. Your job is to present such a clear, professional, and elevated investment opportunity that it becomes harder to say no than yes.

AI gives you the edge through speed, clarity, and positioning power.

2. THE JT FOXX CAPITAL FRAMEWORK

PHASE	DESCRIPTION
P	Position the project for status-aligned investors
R	Raise using automation + psychology
E	Engage with ihgih-impact reporting + AI dashboards
P	Preserve investor lifetime value with repeat offers

3. PHASE 1: POSITIONING THE DEAL LIKE A LUXURY BRAND

Most developers send spreadsheets. You send an investment experience.

AI Tools for Authority Framing:

- Beautiful.ai + ChatGPT-4o → Branded investor decks
- Runway ML → Project sizzle video with voiceover
- Notion/Airtable → Live investor portal

Your pitch materials should reflect:

- Ultra-clean design
- Emotional justification (status, scarcity, exclusivity)
- Logic justification (IRR, equity upside, comparables)

Remember: Investors are human. They buy emotion first and then validate with math.

4. PHASE 2: RAISING CAPITAL WITH AI-DRIVEN SYSTEMS

Never cold-call again.

Use AI to pre-qualify, segment, and automate outreach at scale.

Capital Engine Funnel:

STAGE	TOOL	OUTCOME
Lead Generation	Apollo.io + SparkToro	HNW targeting by interest
Email Sequence	Instantly.ai + GPT	Personalized outreach
Qualification	AI bot (ManyChat or Drift	Filters accredited + affinity investors
Booking	Calendly + Remind AI	Warm call setup
Close	Zoom + JT framing deck	Commitment secured

AI then follows up automatically:
- Sends signed docs
- Triggers wire instructions
- Books onboarding calls

Your team stays lean. The system does the heavy lifting.

5. PHASE 3: ENGAGE INVESTORS LIKE A PRIVATE BANK

Where most developers destroy trust is after the check clears.

Successful developers use investor engagement as a leverage tool to create:
- Future capital raises
- Referral flows
- Reduced scrutiny
- Lifetime investment cycles

AI Investor Portal Includes:
- Project timeline with AI-updated visuals
- Monthly IRR tracking
- Personalized video updates ("InvestorName.mp4")
- Upcoming liquidity options or upgrade offers

Tools:
- Notion + Loom + OpenAI + Airtable
- ChatGPT for monthly investor updates
- Beautiful.ai for quarterly recap decks

6. INVESTOR COMMUNICATION CADENCE (AI-AUGMENTED)

FREQUENCY	ACTION	TOOL
Weekly	Status email with highlights	GPT + Mailchimp
Monthly	Visual report + forecast	Google Data Studio + GPT
Quarterly	Video update + IRR analysis	Loom + Excel AI
Construction Delay	3-day deviation	Alert + alternate vendor suggestion

AI personalization layer adds credibility:
"John, your unit in Phase 1 is 92% complete. You are on track for a 23% IRR."

7. PHASE 4: PRESERVE INVESTOR LIFETIME VALUE
Every raise should be the first of many, not a one-off transaction.

AI Automations That Drive Repeat Investment:
- Trigger new offer sequence 30 days after close
- Send early-access invite to next project
- Automate tax doc delivery + recap video
- Highlight upgraded investment tiers (e.g., preferred returns for multi-deal LPs)

Create a private inner circle:
"Top 12 investors get early equity positions, advisory input, and profit-sharing on brand licensing deals."

This turns passive LPs into loyal, high-conviction backers.

8. CAPITAL STRUCTURES JT USES FOR CONTROL + UPSIDE

STRUCTURE	WHEN TO USE	WHY IT WINS
Preferred Equity	When you want cash without giving up control	Set % return + retain voting rights
Convertible Notes	For fast capital + optional equity	Delay valuation, raise fast
JV Equity Split	When partner brings land/cash	Share profits, keep brand/IP
Profit Participation Bonus	With lifestyle investors	Aligns emotional investors with long-term ROI

AI helps model each structure in minutes and generate investor-specific presentations for clarity and compliance.

9. BUILD THE CAPITAL ENGINE ONCE. USE IT FOREVER.

Most developers rebuild from scratch every deal.

Here is how to build a modular capital machine:

- Replicable pitch stack
- Evergreen funnel + investor pool
- AI-generated updates + reporting templates
- IP that increases the valuation of every future raise

BOTTOM LINE:

Raising capital is not about getting checks. It is about building trust, positioning power, and brand equity that lets you control your deals, not just survive them. With AI, you raise faster, manage smarter, and retain long-term investor velocity.

CHAPTER 18

PROTECTING THE DOWNSIDE— RISK MITIGATION, DEAL STRUCTURING, AND AI-DRIVEN EXIT VISIBILITY

How to de-risk the deal without diluting the upside

In real estate, amateurs chase upside. Successful developers engineer downside protection first, and then scale with confidence.

The game isn't how high it goes. It's how little you lose when it doesn't go perfectly.

This chapter delivers the full risk-control framework, combining brilliant deal structuring, execution safeguards, and AI-driven forecasting to bulletproof your development and protect investor confidence without slowing down momentum.

1. REAL POWER IS IN THE DOWNSIDE MATH

A key rule:

"If the worst-case scenario still makes money, you've earned the right to go big."

Most developers only model the upside. To build:
- Downside models
- Contingency plans
- Exit optionality from Day 1

Risk management isn't fear. It is leverage. It gives you negotiating power, investor trust, and long-term stamina.

2. THE JT RISK STACK: 6 LAYERS OF PROTECTION

LAYER	FUNCTION
1	Land and acquisition arbitrage
2	Phased construction and pre-sales
3	Exit flexibility
4	Contractual insulation
5	Cost-control systems
6	AI forecasting and alerts

Each layer is intentionally engineered into the deal, not tacked on later.

3. LAND + ACQUISITION ARBITRAGE: THE FIRST WIN

The best deals are won before the first brick is laid.

Key Tactics:
- Secure land via options, not full cash outlay
- Negotiate entitlement milestones as contract triggers
- Use seller carry or partner land contributions to reduce front-end capital

AI Tools:
- LandTech + Regrid → Underpriced parcels
- Zillow, Crexi AI scraping → Detect undervalued zoning corridors

If the land is cheap and well-zoned, your margin is protected before you spend $1 on construction.

4. PHASED CONSTRUCTION + TIERED PRE-SALES = CASHFLOW ARMOR

You never want to be fully exposed to build risk. Instead, build in phases tied to sales velocity.

PHASE	TRIGGER	BENEFIT
1	Founders Tier Pre-Sold	Look in IRR + fund first build stage
2	Tier 2 Launch	Adjust pricing, increase margin
3	Final Release	Ride demand wave, time for scarcity premium

JT Principle: Only build what's already sold or soon to be sold.

AI Support:
- Forecast buyer drop-off rates
- Simulate cash flow by the sales phase
- Recommend ideal unit release schedule based on demand signals

5. EXIT FLEXIBILITY = LIQUIDITY CONTROL

Never build with only one exit.

JT uses "exit stacking": 3–5 paths to liquidity baked into the deal.

OPTION	WHEN TO USE
Unit Sales	When emotional buyers dominate
Long-Term Hold + Lease	When cap rates are favorable
Bulk Portfolio Sale	When branding multiplies value
JV Equity Buyout	When partners want a clean exit
REIT Exit	When cashflow history is strong

AI Forecasting Tools:
- Monte Carlo simulations for exit outcomes
- Sensitivity analysis (construction delays, rate hikes, buyer drop-off)

6. CONTRACTS THAT CONTROL CHAOS

Use legal structure to trap upside and deflect risk.

JT's tactics:
- Force majeure clauses with supplier penalties
- Construction bonus milestones tied to time/cost
- Investor equity tied to completion stages, not just capital
- Marketing opt-outs if sales velocity stalls

Tools:
- Docracy + ChatGPT to draft clauses
- Ironclad or Juro for contract workflow automation

A strong contract gives you time, leverage, and options under pressure.

7. AI-POWERED COST CONTROL + EARLY WARNING

Use AI to monitor for cost overruns, contractor issues, or inflationary pressure.

Systems:
- OpenSpace AI → Visual progress versus schedule
- Procore + GPT → Compare estimates versus actuals
- ChatGPT-trained model → Weekly deviation reporting + recommendation engine

Alerts to Set:
- Material cost spike > 8%
- Construction milestone delay > 3 days
- Labor cost overrun > $20K

AI doesn't replace your GC—but it does keep them honest.

8. INVESTOR CONFIDENCE THROUGH TRANSPARENCY

The best risk control with capital partners: over-communicate.

What JT Foxx uses:

- Real-time dashboards (Airtable + Data Studio)
- Monthly video updates (Loom + branded decks)
- "If/Then" scenario trees for all major risks

AI will:

- Auto-generate project health summaries
- Simulate scenarios with investor-facing visuals
- Send alerts if IRR projections shift by more than 10%

Perception of risk = actual risk in investors' minds. AI eliminates uncertainty.

9. REPUTATION = ULTIMATE RISK SHIELD

If you build a reputation for:

- Never missing investor comments
- Always having exit options
- Protecting downside like a fiduciary

… you will raise capital faster and negotiate stronger JV terms, forever.

Use AI to protect reputation:

- Flag missed deadlines
- Auto-escalate buyer concerns
- Trigger proactive investor updates

Reputation compounds. So does negligence. JT chooses predictability over perfection—because predictability wins.

BOTTOM LINE:

JT Foxx doesn't scale with optimism. He scales with control.

Every development has risk. The smart ones model it, pre-negotiate it, and automate detection before damage happens.

If your downside is locked, your upside becomes a weapon. That's what separates project chasers from empire builders.

CHAPTER 19

THE REPLICATION ENGINE — HOW TO DUPLICATE SUCCESS ACROSS CITIES, TEAMS, AND INVESTORS WITH AI PRECISION

From one project to a global portfolio
without burnout, complexity, or team bloat

Growth without replication is just chaos. Successful developers do not scale by doing more. They scale by creating duplicatable infrastructure that multiplies success with precision, speed, and brand control.

This chapter outlines the full Replication Engine—an AI-powered framework to duplicate your winning development model across new cities, new partners, and new investor pools without losing quality, control, or positioning.

1. SUCCESS IS NOT RANDOM. IT'S ENGINEERED.

One-off success is luck. Engineered replication is leverage.

Successful developers do not chase bigger deals. They build a controlled system that allows others to build on their playbook while they retain the brand, the process, and the profit trigger points.

The Replication Engine gives you:
- Operational clarity
- Partner duplication

- Brand protection
- Asset-light expansion

2. THE REPLICATION MODEL

PHASE	DESCRIPTION
C	Codify your playbook (systemization)
A	Automate execution (AI tools)
S	Syndicate partners (replicable capital + JV structure
E	Expand under one brand (licensing + territory rights)

Now you can go from a project builder to a portfolio architect without adding team or time drag.

3. CODIFY YOUR DEVELOPMENT IP ONCE & SELL IT FOREVER

First, extract and document your full development process, including:

- Site selection criteria
- Zoning battle playbooks
- Capital stack structure
- Sales funnel sequences
- Marketing and media templates
- Construction vendor roles + red flags
- Buyer persona + emotional triggers
- Post-close engagement model

Use Notion, Loom, and ChatGPT to turn this into:

- Standard operating procedures (SOPs)
- Checklists
- Templates
- Scripts
- Explainer videos

It is not administrative work. It is your future licensing model.

4. AUTOMATE EXECUTION WITH AI ASSISTANTS

Once codified, every function becomes a module that AI can either run or manage.

MODULE	AI TOOL	OUTCOME
Land scouting	Regric + ChatGPT prompt stacks	Auto-suggests qualified parcels
Capital raising	Instantly.ai + CRM	Pre-built investor funnel
Zoning naviga-tion	AI bot trained on city code	Accelerates permit prep.
Sales scripts	ChatGPT custom personas	Tailored pitch by avatar
Investor reporting	GPT + Airtable	Monthly updates auto-drafted
Event promo-tion	Canva + Jasper + Mailchimp	Branded campaign rollout

This turns your process into a franchise-style machine.

5. SYNDICATE PARTNERS: THE JT JV MODEL

Instead of expanding alone, you create a bench of qualified partners who:
- Bring capital or land
- Follow your model
- Get limited territory rights
- Pay a percentage of revenue, licensing, or profit-share

Your system does the heavy lifting. They execute.

Structure:
- 51% brand + system control
- 49% local execution, upside, and labor

Include:
- Brand license
- Execution compliance terms
- Exclusive territory windows (with performance clauses)

AI automates partner onboarding, training, and deal tracking.

6. BRAND PROTECTION ACROSS CITIES

Scaling doesn't work without brand consistency.

Key Rules:
- Centralized asset library (logos, pitch decks, design standards)
- Locked pricing tiers per phase (no undercutting)
- Shared testimonials, investor NPS, and press mentions
- AI-monitored brand compliance (social media scraping + audits)

Tools:
- Brandfolder + Notion
- ChatGPT brand tone checker
- Grammarly + Surfer SEO for AI-written consistency

The brand is the moat. Treat it as an asset class.

7. TERRITORY EXPANSION: WHO BUILDS WHERE AND HOW

License strategically:
- 1–2 partners per major metro
- 1 country manager per international region
- Minimum quota: 1 project per 12 months
- Revocable license if underperformance

Each market must follow:
- Brand architecture
- Capital structure
- Lifestyle branding model
- Exit strategies

Use Airtable + AI CRM to track licensing rights, deal velocity, and partner metrics.

8. GLOBAL REPLICATION PLAYBOOK DEPLOYMENT

Your global expansion folder includes:

ASSET	DESCRIPTION
Investor pitch stack	Editable slides + scripts
Deal structure matrix	Pick-a-model template (Equity/Pref/License)
Project tracker	Timeline template + AI alerts
event sequence	Pre/post-sell event playbook
Media kit	Templates for press, influencer, and PR
Dashboard shell	Airtable/Notion master copy

Each partner receives a turnkey replication kit, backed by your real-world data.

9. THE INVISIBLE EMPIRE: ASSET-LIGHT, BRAND-HEAVY

You are now operating a development portfolio that:
- You don't fully fund
- You don't fully manage
- You don't fully build

But you fully profit from:
- AI ensures control. Licensing ensures spread.
- The result: maximum upside, minimal exposure.
- You've built a machine, not a hustle.

BOTTOM LINE:

Most developers scale chaos. You scale control.

The Replication Engine gives you a duplicatable, defensible, and highly profitable business model—with global reach and brand equity baked in.

CHAPTER 20

FROM NICHE TO EMPIRE — THE 7-FIGURE LICENSING PLAYBOOK FOR SPECIALIZED REAL ESTATE BRANDS

How to monetize your model without
building more or managing more

If you're still building deal after deal, you're working too hard. You don't scale by stacking projects—JT scales by stacking licenses.

When you build a specialized real estate brand with strong emotional and financial appeal, the next step isn't more labor; it's leveraging your brand, process, and results into a monetizable licensing empire.

This chapter gives you the 7-Figure Licensing Playbook: how to package your development model, sell it to qualified partners, and turn one winning concept into dozens of high-margin, low-overhead revenue streams.

1. LICENSING VERSUS FRANCHISING IS CONTROL WITHOUT REGULATION

Franchising = regulatory maze, legal bottlenecks, corporate complexity

Licensing = fast, lean, enforceable, and cash-rich

We use licensing to:

- Retain brand control
- Monetize processes without operational drag

- Expand globally while staying asset-light
- Structure high-profit recurring income with zero project exposure

The brand becomes the business.

2. THE JT FOXX LICENSING STACK

STEP	DESCRIPTION
B	Build a market-proven development model
R	Replicate once to prove it's duplicatable
A	Assemble the license package
N	Negotiate eclusive territory deals
D	Deliver systems with performance enforcement

3. STEP 1: BUILD THE PROVEN MODEL

Before licensing, you need:

- A specialized development model (e.g., auto condominiums, hangar homes, branded storage, etc.)
- At least one high-margin, high-sellout project
- Clear emotional buying triggers and positioning advantages
- AI-powered system for lead generation, sales, capital, and reporting

Proof of results and process clarity is what buyers are really licensing.

4. STEP 2: REPLICATE ONCE TO PROVE IT'S SCALABLE

Build the second project in a different city or with a JV partner to demonstrate:

- System portability
- Brand value in other markets
- Repeat buyer behavior
- Margin retention

This now becomes your case study to command premium license fees.

5. STEP 3: ASSEMBLE THE LICENSE PACKAGE

Your licensing kit must include:

ASSET	PURPOSE
Brand Guide	Visuals, tone, logo rules, messaging
Development SOPs	Full Notion playbook of steps, roles, checklists
Financial Models	Proformas, IRR matrixes, deal structures
Sales + Marketing Funnels	Copy, CRM templates, AI chat flows
Investor Pitch Asserts	Decks, sizzle video, objection handling scripts
Contract Templates	Legal structures, JV docs, HOA framework
AI Dashboard Shell	Airtable/Notion-based command center
Training Portal	Loom/Thinkific modules, system walk-throughs

The licensee is buying certainty. Package everything they need to win.

6. STEP 4: NEGOTIATE EXCLUSIVE TERRITORIES WITH SCARCITY FRAMING

Position your licenses like premium real estate: "We're offering one license per city or region. First in owns the territory."

Structure your deal with:
- $50K–$250K upfront license fee
- 5–8% gross revenue royalty
- Optional consulting support retainer
- Annual renewal clause tied to performance

- Revocable rights if the brand is misused or results underdeliver

Licensees are paying for speed, credibility, and protection against competition.

7. STEP 5: DELIVER + ENFORCE EXECUTION STANDARDS

This is where most people fail. JT ensures every licensee:
- Follows the model
- Protects the brand
- Hits performance benchmarks

Tools:

- Brand audits (monthly AI check of marketing, design, messaging)
- Sales + investor key performance indicator (KPI) reporting (dashboard-linked)
- Secret shopper lead test (AI-based response analysis)
- AI-powered onboarding flow (automated training + task sequence)

When brand equity is consistent, you multiply valuation.

8. AI = THE LICENSING LEVERAGE MULTIPLIER

You don't need a team of trainers or account managers. AI does the heavy lifting.

LICENSING FUNCTION	AI AUTOMATION
Training Delivery	GPT + Thinkific auto-modules
Lead Capture	Custom chatbot with territory-specific funnels
Email Sequences	AI-personalized investor + buyer communications
Onboarding Compliance	Airtable checklists + AI status alerts
Performance Tracking	Data Studio dashboards with GPT insights

With AI, every licensee runs like a well-trained internal team.

9. GLOBAL EXPANSION STRATEGY (THE JT ROLL-OUT MAP)

Start licensing in:

- Tier 2 U.S. cities with wealth + collector culture
- Expat hubs with UHNW buyers (Dubai, Singapore, Cape Town, Monaco)
- Regions with low developer sophistication but high buyer demand

Roll out with:

- One elite licensee per metro
- Group mastermind (monthly AI-supported Zoom)
- Annual in-person retreat (event monetized)
- Licensee NPS tracking system for brand quality control

You're now scaling an elite club, not a development business.

10. MONETIZATION MULTIPLIERS

Licensing isn't just upfront cash. JT stacks revenue:

SOURCE	DESCRIPTION
Upfront Fee	$50K-$250K /license
Royalty	% of gross revenue or profit
Consulting	Paid guidance on key phases
Events	Paid access for licensees + prospects
Brand Sponsorships	Partner brands across all locations
Mastermind	Premium upsell for licensees + inner circle
Portfolio Buyout	Consolidated roll-up value

Licensing builds recurring revenue, upside equity, and exits without building anything else.

BOTTOM LINE:

You can chase the next project, or you can license the last one—again and again.

If your model is proven, your process is documented, and your brand is tight, you don't need more effort. You need more leverage.

Licensing creates an empire without overhead. Scale without stress.

CHAPTER 21

THE LEGACY LAYER — TURNING YOUR SPECIALIZED REAL ESTATE BRAND INTO A LONG-TERM WEALTH ENGINE

From high-income projects to
multi-generational, asset-backed influence

1. MOST DEVELOPERS BUILD DEALS.

2. SUCCESSFUL DEVELOPERS BUILD LEGACY.

3. DEALS EXPIRE. BRANDS COMPOUND.

Legacy isn't just about wealth. It's about how you built it, who benefits from it, and what continues when you don't show up.

This chapter reveals how to embed legacy into your real estate empire—from structure to succession, from media to family planning, so that your specialized development brand becomes a long-term vehicle for cash flow, control, and generational relevance.

1. THE END GOAL IS NOT FREEDOM — IT'S CONTINUITY

High-income is a trap if your empire depends on you. Successful developers do not just exit deals. They engineer asset-backed continuity, where:

- Cash flow is protected
- Brand equity outlives them
- Wealth is structured to benefit family, clients, and causes—on their terms

Legacy is built with intentional architecture, not emotional optimism.

2. BUILD THE BRAND TO OUTLIVE YOU

- A name is not a brand
- A positioned identity is

Create a brand that:

- Represents values, not a person
- Embeds core frameworks, not just charisma
- Can be licensed, sold, or stewarded without dilution

Tools:

- Brand Bible (visuals, mission, tone, philosophy)
- SOP Vault (Notion or AirManual)
- AI-trained brand tone checker
- Trademark protections across products, logos, and taglines

If your brand can't scale without you, it's not a legacy; it's a job.

3. FROM FOUNDER INCOME TO PERPETUAL IP MONETIZATION

Structure your income for control even when you're no longer operating.

Licensing + royalty models:

- 8-10% recurring on gross revenue from licensees
- Equity stakes in licensee growth roll-ups
- Content licensing (training, media, coaching rights)
- Digital products powered by AI agents and evergreen delivery

Add:

- Trademarked products
- Licensing trust entities
- Estate-protected cashflow systems

Your intellectual property becomes the annuity.

4. HOLD REAL ESTATE LIKE A FAMILY OFFICE, NOT A DEVELOPER

Stop flipping your best assets. Hold them under structures that protect, appreciate, and pass through seamlessly.

Successful developers' Framework:

LAYER	STRUCTURE
Legal	LLC per project, rolled into master holding company
Tax	Series LLC or LP with flow-through benefits
Trust	Irrevocable trust with brand + IP ownership
Ops	Virtual family office with an AI dashboard for heirs
Exit	Trigger-based events (sale, refinance, license bundle)

Tools:
- Carta + Notion for ownership mapping
- AI-powered portfolio dashboards for family education

5. TRAIN THE NEXT GENERATION LIKE A BOARDROOM, NOT A CLASSROOM

Your kids, heirs, or successors don't need an inheritance. They need financial fluency and operational exposure.

Successful Developers Approach:
- Shadow deals from Day 1
- Assign controlled equity with vesting triggers
- Use AI dashboards to walk them through deal dynamics
- Create gamified incentives for KPI milestones

Teach:
- Deal structure
- Brand protection
- Licensing leverage
- Exit timing and capital preservation

You don't raise heirs. You build future partners.

6. MEDIA LEGACY: DOCUMENT EVERYTHING

If you're not building your legacy in public, you are relying on memory.

Document:
- Your principles (book, podcast, video series)
- Your process (SOPs, client case studies)
- Your personal evolution (video logs, AI-narrated recaps)

Tools:
- Runway ML for legacy storytelling
- ElevenLabs for voice cloning
- ChatGPT to auto-document lessons from meetings, deals, and events

This becomes your media vault—a digital asset passed down or packaged into public legacy content.

7. LEGACY CAPITAL STRATEGY: EXIT WITHOUT SELLING OUT

Never let urgency rob you of optionality. Successful developers build with strategic patience by embedding legacy-minded exits:

OPTION	TRIGGER	OUTCOME
Sale to Portfolio Buyer	3+ licensees, $5M + EBITDA	Brand roll-up exit
Private Equity Stake Sale	$2M+ net profit, replicable model	Partial cash-out, keep control

OPTION	TRIGGER	OUTCOME
Brand Equity IPO	High brand awareness, global demand	Exit + investor buy-in
Intergenerational Transfer	Wealth plan triggers	Family or foundation continuity

All built on one principle: You own the game. You write the end.

8. THE FINAL WEALTH FORMULA

Here is the full architecture successful developers use to build legacy:

Specialized Asset + Branded Position + AI Execution + Scalable Licensing + Wealth Structure = Legacy Empire

- This is not just about money. It is about identity, influence, and permanence.

BOTTOM LINE:

Projects end. Capital dries up. Markets shift. But legacy, if built intentionally, transcends it all.

You are not just a developer. You are a founder of a movement. The brand, the systems, the deals, the licenses—they are just vehicles. Legacy is the destination.

CONCLUSION: AI STRATEGIES FOR PROFITABLE EXECUTION IN SPECIALIZED REAL ESTATE DEVELOPMENT.

Your Blueprint. Now Execute Like It!

You've just been handed the entire AI-powered playbook to dominate niche real estate development.

This is not motivational. It is a system.

A proven model for building:

- Specialized, emotionally-driven real estate assets
- Investor-ready capital machines
- AI-integrated execution frameworks
- Brand platforms that scale, license, and leave legacy

This is how JT Foxx and William Böll build differently. And now you can too.

YOUR NEXT MOVES, IF YOU'RE SERIOUS:

1. Pick a Niche. Own It.

Auto condominiums. Hangar homes. Luxury storage. Branded Short Term Rentals (STRs).

Don't dabble. Dominate. Niche positioning is what gives you margin, speed, and media leverage.

When you own the category, you set the price.

2. Systemize With AI.

Execution intelligence beats hustle.

Every section of this book showed you how to use AI to:
- Scout land
- Underwrite deals
- Build funnels
- Raise capital
- Manage builds
- Report to investors
- Replicate success

Use the AI stack, or be replaced by it.

3. License Your Model.

You don't need more projects. You need more people executing your system, under your brand, while you collect.

Licensing = wealth without repetition.

4. Build a Brand, Not a Business.

A brand multiplies every dollar, protects every asset, and compounds every exit.

Your development isn't merely square footage; it is a media entity, a tribe, and a prestige engine.

Make people want in. Then charge for access.

5. Think Like a 9-Figure Founder. Be part of the 3-comma club.
- You're not a builder.
- You're not a developer.

- You are a platform architect with AI-enhanced power and IP-backed control.

FINAL WORDS FROM WILLIAM BÖLL:

Real estate will always be one of the greatest wealth vehicles on earth. But how you play the game determines what you keep. This book didn't teach you how to play. It gave you the board, the rules, and the power tools.

Now build something the market can't ignore. Then scale it so big, they can't catch up. Then license it so you never have to again.

AI 2.0 —
THE FUTURE OF AUTONOMOUS REAL ESTATE TEAMS & GLOBAL MICRO-ASSETS

Scale with no office, no employees,
and no borders

The next phase of real estate won't be owned by the biggest developers. It will be led by the leanest, smartest, and most autonomous operators.

AI 1.0 was about optimization. AI 2.0 is about orchestration.

This bonus chapter gives you a glimpse into what JT Foxx is already preparing for: a fully autonomous, borderless, high-margin real estate model powered by AI teams and global asset leverage.

1. THE AI TEAM STACK — NO EMPLOYEES. PURE EXECUTION.

You don't need a staff. You need an AI-powered execution unit.

AI Team Template:

FUNCTION	AI ROLE	TOOL(S)
Deal Scout	Identifies unerpriced land/assets	LandTech + ChatGPT agents
Underwriting Analyst	Builds pro formas + sensitivity models	GPT-4o + Excal AI
Capital Manager	Runs outreach, emails, follow-up	Instantly.ai + GPT
Construction Monitor	Tracks progress via camera + docs	OpenSpace + GPT Summary
Buyer Concierge	Handles inbound buyer questions	AI chatbot + cRM triggers
Media Producer	Cuts reels, writes posts, generates ads	Runway ML + Jasper + Canva
Investor Reporter	Auto-generates reports, videos, updates	Airtable + GPT Dashboards

Each "employee" runs 24/7, never complains, and scales instantly.

2. MICRO-ASSETS: THE RISE OF HIGH-MARGIN, LOW-OVERHEAD PLAYS

Traditional development = millions in capital, years of exposure. The new model = micro-assets with macro leverage.

Examples:

- Auto condominiums (12–30 units)
- Hangar homes (for private airports)
- Collector storage + lounge hybrids
- Branded STR enclaves
- Luxury modular "identity units" in global hotspots

Why they win:
- Built fast
- Emotionally-driven buyers
- AI can operate end-to-end
- Resale = prestige product, not just square feet
- Can be licensed, not just sold

The margins go up as the scale goes down, if positioned right.

3. NO HQ NEEDED — RUN YOUR PORTFOLIO FROM ANYWHERE

With AI 2.0, or above, the principle is simple:

"Where you live is irrelevant. What you control is everything."

How to virtualize:
- Use Notion or Airtable as your central command center
- Integrate with GPT-4o for daily summaries, alerts, and recaps
- Add Zapier to automate tasks across tools
- Deploy AI chatbots as buyer agents, investor liaisons, and operations leads

Your laptop or your smartphone becomes your empire.

4. AI-TRAINED LICENSING REPS — SELL YOUR SYSTEM WHILE YOU SLEEP

Imagine:
- 100+ inbound licensee leads
- AI qualifies them, answers objections, and sends decks
- Books a live Zoom or closes with Stripe + DocuSign

Licensing AI Setup:
- Website funnel → Chatbot + lead scoring
- Licensee CRM → Tiered access to pricing, territory maps, testimonials

- Closing flow → AI-narrated proposal + payment + onboarding portal

This is not future tech. It is current leverage.

5. THE GLOBAL EXPANSION MODEL: MICRO ASSETS, LOCAL OPERATORS

You don't build everywhere. You license, JV, or fund operators who run your model locally.

Framework:

REGION	ASSET TYPE	PARTNER TYPE
Dubai	Auto condominiums	Local UHNW with land
Texas	Ranch STR enclaves	Developer with build team
Monaco	Collector vault + lounge	Luxury brand collab
Singapore	Vertical car condos	Tech-backed RE fund

You bring:
- The brand
- The system
- The playbook
- The AI tools

They bring:
- Land
- Build team
- Execution energy

6. EXIT THROUGH IP, NOT JUST REAL ESTATE

Wealth doesn't come from selling buildings. It comes from selling systems, brands, and IP with built-in proof.

You can now:
- Sell your licensed brand portfolio to a roll-up

- Sell the tech layer that powers your development engine
- Sell digital-only licenses for markets you'll never touch
- Create AI-powered investment portals for UHNW buyers

You are no longer just a developer. You're a platform founder in the real estate space.

BOTTOM LINE:

AI 2.0 and above isn't coming. It's here.

If you build now with:

- Autonomous teams
- Systemized assets
- Brand-layered plays
- Global licensing and AI leverage

You're not scaling real estate. You're scaling influence, control, and freedom—on your terms.

JT Foxx is already executing this.

Now make it your advantage.

APPENDIX

TOOLS, TEMPLATES, AND PROMPTS FOR EXECUTION-READY DEVELOPERS

Specialized real estate development integrates location strategy, tailored design, and unique amenities to meet the needs of a niche market segment. It combines financial structuring, regulatory navigation, and brand positioning to deliver differentiated projects with premium investor and end-user value.

This collection brings together the core execution tools and case studies that define our advantage in specialized real estate development. Each component is designed to demonstrate not only the sophistication of our AI-enabled systems but also their direct application to capital raising, investor trust, and profitable delivery. The main components of the core execution tools are:

- JT Approved Tech Stack — A curated set of tools vetted through real-world application and JT Foxx's execution standards. This stack ensures developers and investors operate with clarity, precision, and speed rather than guesswork.
- AI-Powered Capital — Frameworks for structuring, raising, and managing capital using artificial intelligence to optimize investor targeting, deal framing, and perpetual fundraising momentum.
- AI Tech Stack Cheat Sheet — A condensed, actionable guide to the essential AI tools powering site

selection, financial modeling, branding, investor reporting, and exit planning. Built for rapid adoption and replication across projects.

- Velocity Prestigious — A live case study of the Velocity Prestigious Auto Residences—our flagship luxury auto condominium brand. It demonstrates how identity-driven real estate, backed by AI systems, creates trophy assets with replicable global potential.
- PSC Script (Pitch–Sell–Close) — The AI-enhanced sales framework that compresses sales cycles to 21 days or less. This proven script system aligns with identity-driven buyers, filters serious prospects, and maximizes close rates.
- Sample Pro Forma—A model of financial performance designed to illustrate the profitability of specialized developments. It showcases how capital stack engineering, AI-driven forecasting, and brand leverage deliver both investor protection and outsized returns.

Together, these materials form a complete playbook: technology stack + capital strategy + case execution + sales methodology + financial validation. A specialized real estate development playbook is built on four interdependent subcomponents. Market Intelligence & Positioning defines the target niche, demand drivers, and competitive differentiation; Design & Operations focuses on tailored architecture, amenities, and operating models; Capital Stack & Financial Engineering structures debt, equity, and investor protections for optimal returns; and Technology & Execution Frameworks integrate digital tools, data analytics, and disciplined project delivery to ensure scalability, transparency, and investor confidence. By integrating the subcomponents of the playbook, it ensures projects move from concept to profitable reality with clarity and investor confidence. Below are the necessary subcomponents of a successful playbook:

A. JT-Approved Tech Strack for Real Estate Execution
B. AI-Powered Capital Raising & Syndication
C. Velocity Investor Binder
D. Pitch-Sell-Close (PSC) Script
E. Sample Pro Forma from AI
F. AI-Generated Design of Auto Condominiums
G. Final Takeaway

This is not fluff. It's the exact tactical layer you need to execute fast, lean, and with clarity—aligned 100% with JT Foxx's systems and results-driven mindset. This appendix isn't for reading. It's for building. Every AI tool, structure, and template here is designed to compress time, multiply leverage, and scale profit with precision. Execute relentlessly. License intentionally. Exit intelligently.

A. JT-APPROVED TECH STACK FOR REAL ESTATE EXECUTION

FUNCTION	TOOL(S)	PURPOSE
Land Scouting	LandTech, Regric, Chat GPT	Identify undervalued, zoned parcels with AI filters
Deal Modeling	GPT-4o, Excel AI, Causal	Run pro formas, IRR models, downside stress tests
Capital Raising	Instantly.ai, Apollo.io, Pipedrive + GPT	Investor sourcing, nurturing, automated pitch sequences
Investor Reporting	Airtable, Notion, Data Studio, Loom + GPT	Real-time dashboards, updates, personalized communications
Construction Oversight	OpenSpace.ai, Procore, CoConstruct	Visual progress, task tracking, GC accountability

FUNCTION	TOOL(S)	PURPOSE
Buyer Funnels	ManyChat, High-Level, Unbounce, GOT chat flows	AI qualification, emotional triggers, follow-up at scale

A1. LICENSING PLAYBOOK TEMPLATE (JT FOXX FRAMEWORK)

- Core License Package:
- Territory rights (defined by ZIP/country/region)
- Brand usage (logo, pitch decks, marketing assets)
- Development SOPs (Notion-based)
- Financial models (IRR matrix, investor return scenarios)
- Sales funnels + CRM template
- AI dashboard portal (Notion + Airtable shell)
- Access to an AI-trained GPT assistant
- Quarterly brand audit + mastermind invite

Key Deal Terms:
- $50K–$250K upfront license fee
- 6–10% royalty on gross revenue
- 12-month minimum execution clause
- Performance-based renewal
- Optional consulting retainer available

A2. AI PROMPT STACK BY DEPARTMENT

For Deal Analysis:
"Using these inputs [land price, build cost, sale price per sq. ft.], calculate IRR, profit margin, and breakeven scenarios under 3 risk levels."

For Investor Pitching:
"Write a high-emotion investor email for a premium auto condo development in [City], highlighting scarcity, passion-driven returns, and founder track record."

For Buyer Chatbot:
"You're an expert concierge for a luxury auto condo brand. Qualify leads by asking about car collections, investment goals, and customization interests."

For Monthly Updates:
"Summarize project milestones for this month. Include IRR forecast, construction status, investor note, and next 30-day plan. Use a tone of authority + clarity."

For Licensing Recruitment:
"Draft a landing page offer for licensing our real estate development brand. Emphasize exclusive territories, turnkey execution systems, and emotional asset class growth."

A3. NOTION/AIRTABLE ARCHITECTURE (SAMPLE)

Notion Dashboard:
- SOP Hub: Sales, Capital, Build, Marketing, Exit
- Licensee Portal: Docs, training, progress tracking
- Brand Guidelines: Visuals, language, IP policies
- Team AI Log: Chat summaries, decisions, and prompts used

Airtable Dashboard:
Investor CRM (funding stages, communications status)
- Project Tracker (unit status, lead flow, revenue)
- Licensee Map (territory, revenue, compliance status)
- KPI Metrics (daily + monthly AI-pulled reports)

A4. EXIT PLANNING TOOLKIT

EXIT PATH	TRIGGER	TOOLS TO PREP
Portfolio Sale	$5M+ EBITDA across 3+ projects	CIM doc (Beautiful.ai), consolidated dashboard, brand equity proof
Licensing Roll-up	10+ licensees across 3 regions	SOP suite, royalty tracker, legal docs, media kit

EXIT PATH	TRIGGER	TOOLS TO PREP
JV Buyout	Passive investor request	Contract templates, AI valuation report, investor ROI history
Family Transfer	Wealth preservation milestone	Trust setup, IP ownership structure, family ops dashboard

A5. JT FOXX'S EXECUTION FILTERS (USE BEFORE ANY MAJOR DECISION)

- Does this move increase control or dilute it?
- Can this system run without me in 90 days?
- Will this build long-term brand equity or temporary income?
- Is the emotion driving the buyer or the math?
- Can this be duplicated, licensed, or exited—without me rebuilding it?

B. AI-POWERED CAPITAL RAISING & SYNDICATION

Here's the complete package for AI-Powered Capital Raising & Syndication tailored to your specialized real estate development project (e.g., using the author's Velocity Prestigious Auto Residences and Velocity Performance Alliance projects). This includes:

LEAD NAME	TYPE	WARMTH SCORE	RECENT ACTION	SENTIMENT	RECOMMENDED NEXT ACTION	VALUE
John T., Monaco	Family	92	Viewed deck, asked for model	Positive	Send term sheet + concierge call	$500K
Natalia C., Austin	Angel	65	Clicked landing page	Neutral	Retarget w/ prestige benefits	$250K

| Raj S., Dubai | HNWI | 78 | Opened PPM | Mixed | AI-generated check-in email | $1M |
| Velocity Syndicate | Syndicate | 88 | Joined Data Room | Positive | Invite to lead tranche | $2M |

Investor CRM Dashboard Mockup — AI-Augmented

TITLE: VELOCITY INVESTOR INTELLIGENCE CONSOLE

Features Powered by AI:

- Auto-tag investor type (Angel, Family Office, Syndicate)
- Predictive Warmth Score (based on behavior patterns)
- Sentiment Analysis (NLP on all email/call logs)
- Next-Best Action Engine (GPT recommends what to do)

B1. CUSTOM AI-ENHANCED PITCH DECK FLOW

"Framed for Speed, Engineered to Convert"

- Cover Slide – Bold image of auto condo + tagline ("Engineered for Passion. Designed for Security")
- Market Opportunity – $23B+ global collector garage demand (AI-enhanced comps by region)
- Problem – HNW owners lack lifestyle-secure car storage with brand cachet
- Solution – Branded luxury auto condominiums + private club experience
- Brand Architecture – Velocity: global expansion potential (Dubai, Monaco, Scottsdale)
- Investment Summary (Dynamic AI Slide) – Personalized to viewer (e.g., "Family Office-Fit Structure")
- Capital Stack Model – AI-modeled waterfall with IRR, pref return, and tokenization option

- Syndication Strategy – Multi-tier LP structure + real asset tokenization (AI-managed onboarding)
- AI Tech Stack – AI in site selection, investor onboarding, digital twins for design
- Exit Pathways – Strategic REIT acquisition or full brand/IP exit
- Call to Action – Book a private "call" or "Join the Investor Room"

Syndication Playbook: AI Execution Blueprint

"RAISE WITH PRECISION. SYNDICATE WITH CONFIDENCE."

- Step 1: Investor Persona Engine
- Feed previous LP data + CRM lists into an AI tool like Clay or Affinity
- Create investor archetypes (e.g., "Retired NASCAR executive," "Ex-tech founders," "Wine+Car collectors")
- Step 2: AI-Based Outreach
- Use GPT-4-generated personalized email copy
- Auto-adjust tone and content based on type (family office versus syndicate versus retail)
- Step 3: Tiered Syndication Build
- Tier 1: Anchor LP (target $1M+ single investor)
- Tier 2: Co-GP Syndicate (AI-manage K-1s, waterfall)
- Tier 3: Tokenized LP Pool (optional) – US + UAE accredited backers
- Step 4: Digital Data Room + Bot Assistant
- Build a GPT-powered assistant inside your data room to answer:
- "What is the preferred return?"
- "When does the first distribution start?"
- "What is the land basis and IRR?"
- Step 5: Smart Follow-Up
- NLP-based analysis of investor replies via Superhuman or Front

- Automate: retargeting email + case study + private invite to tour/demo

B2. CAPITAL STACK OPTIMIZATION WITH AI TOKEN MODELING

AI played a critical role in shaping a three-tiered tokenized capital stack:

- Class A (Yield First): 8% preferred return, 2% IRR floor, modeled to attract stable capital.
- Class B (Revenue Participation): Modeled waterfall triggers based on phase-level absorption velocity.
- Class C (Developer Incentive): AI assessed IRR scenarios (17–24%) to time unlock thresholds and mitigate dilution.
- AI simulations ran 10,000+ capital event scenarios to ensure all downside protections were embedded in smart contracts.

B3. AI TECH STACK CHEAT SHEET — CAPITAL RAISE EDITION

FUNCTION	TOOL/STACK	DESCRIPTION
Investor Prospecting	Clay + Apollo + Affinity AI	Scrape, analyze, and rank top investor matches
Personalized Outreach	Lavender + GPT-4o	Auto-customize email intros & follow-ups
Data Room + Bot	DocSend + GPT Assistant	24/7 doc Q&A powered by your PPM & deck
CRM Intelligence	HubSpot + OpenAI Plugin	Investor timeline + next best action modeling

Syndication Legal	Carta + Tokeny	Automate cap table, tokenized shares (optional)
Analytics	Mutiny + Hotjar + GA4	Behavior-based optimization of portal and decks
Compliance	Persona + alloy	AI-powered know your client (KYC), anti-money laundering (AML), and accreditation checks

4. VELOCITY INVESTOR BINDER

Velocity Prestigious Auto Residences
Investor Binder
AI-Powered Capital Raising & Syndication Edition
Private & Confidential

Velocity Performance Alliance – Investor Binder Summary

1. CORE INVESTMENT THESIS

Velocity Performance Alliance (VPA) is pioneering a new category of branded real estate: luxury automotive lifestyle vaults (auto condominiums). These are not speculative builds; they are scarcity-driven, brand-anchored assets engineered for high margins, fast absorption, and institutional exit premiums.

Market Leadership: First-to-market in Scottsdale/Phoenix metro with a 64-unit development (128,000 SF).

Global Replicability: A portfolio play targeting Miami, Monaco, Dubai, Austin, Las Vegas, and Singapore.

Investor Alignment: LP-first returns, conservative leverage, and multiple exit pathways ensure downside protection with outsized upside.

2. SAMPLE PHASE 1 DEVELOPMENT – SCOTTSDALE / PHOENIX METRO

Total Development Cost (TDC): $46.0M

Total Project Revenue: $62.9M

Gross Profit: $15.5M (32.7% margin)

Investor Equity Raise: $6.9M (minimum entry $250K)

Developer Equity: $2.3M subordinated (absorbs first loss, reinforcing investor protection)

Build Scope:

64 shell units (128,000 SF, $263/sf build cost)

Premium "event-ready" upgrade options

Phased monetization: HOA dues, event rentals, concierge services, resale fund

3. CAPITAL STACK OVERVIEW

The capital stack is designed to balance leverage efficiency with investor security, giving LPs downside protection while maximizing IRR on successful execution.

Senior Debt – $27.6M

60% Loan-to-Cost (LTC)

Interest-only facility to reduce early cash strain

DSCR stress-tested at +300 bps interest rate shocks

Preferred Equity – $9.2M

Fixed 8% coupon

Sits senior to LP equity but junior to debt

Provides structured downside cushion for debt coverage

Investor LP Equity – $6.9M

Priority return of capital

8–10% preferred return (quarterly accrual)

70/30 profit split post-pref (LP/Sponsor)

If IRR exceeds 15%, promote adjusts to 60/40 (still LP-favorable)

Developer Equity – $2.3M

Subordinated "at-risk" capital

Paid only after LPs achieve pref + profit share

Alignment mechanism ensuring sponsor's risk is behind investors

Capital Protection Features:

Minimum 2% IRR clause before any promote

Escrowed reserves for contingency

GMP contracts on construction to reduce cost overruns

Completion guarantees

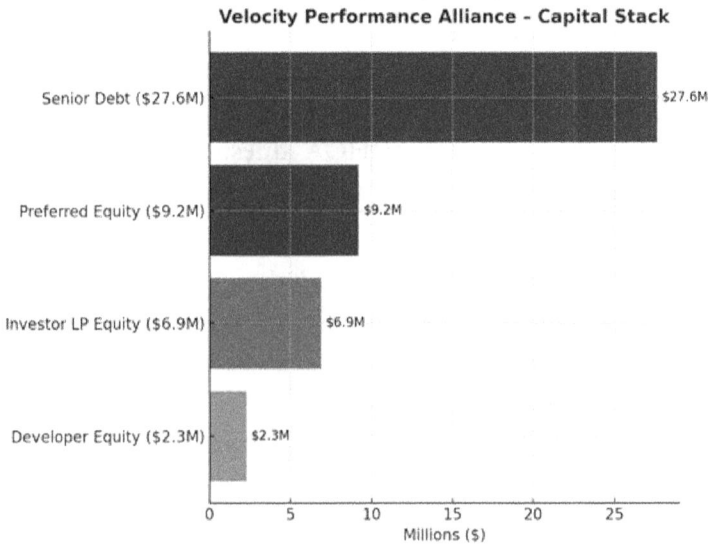

Velocity Performance Alliance - Capital Stack

Senior Debt ($27.6M) — $27.6M
Preferred Equity ($9.2M) — $9.2M
Investor LP Equity ($6.9M) — $6.9M
Developer Equity ($2.3M) — $2.3M

Millions ($)

4. NET EBITDA PROFILE

The project produces meaningful stabilized earnings, creating cash flow optionality and enhancing portfolio valuation for exit:

Stabilized Net EBITDA: $2.56M annually

EBITDA Margin: ~38–40% post-HOA + service fees

Recurring Revenue Sources:

HOA dues and facility fees

Event and filming rentals

Premium concierge services

Club memberships

Exit Multiple Basis:

Valuation at 6–8x EBITDA = $15.4M–$20.5M institutional exit value for Phase 1 alone

Portfolio multiples increase with replication across multiple cities

Key Insight: Unlike speculative condo flips, Velocity vaults generate ongoing EBITDA streams that enhance exit multiples, making them attractive not only as real estate assets but also as brand-anchored, yield-driven businesses.

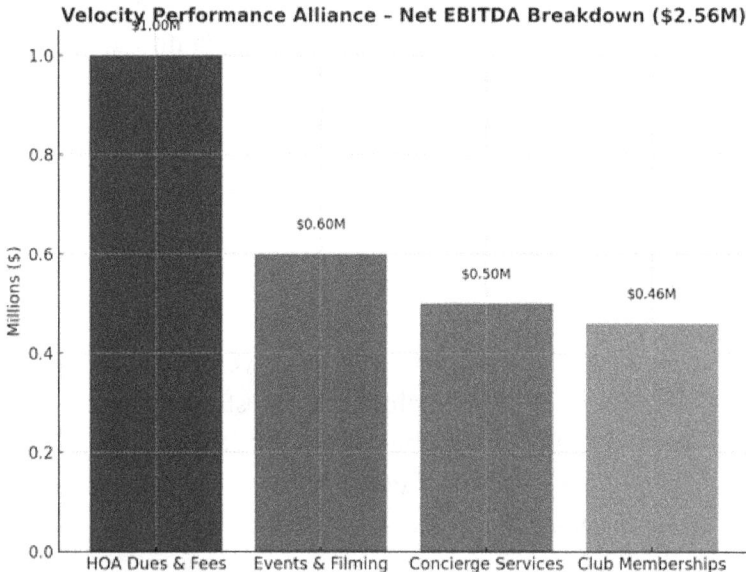

Velocity Performance Alliance - Net EBITDA Breakdown ($2.56M)

5. INVESTOR RETURN PROFILE

Target IRR: 22–26%

Equity Multiple: 1.7x–2.5x depending on hold period

Exit Windows:

24 months: 24% IRR (2.5x multiple)

36 months: 21% IRR (2.0–2.3x multiple)

48 months: 17% IRR (1.8x multiple)

Liquidity Enhancements: Internal resale fund + brand-managed unit buybacks

6. STRATEGIC EXIT OPTIONS

Refinance: Extract equity while maintaining brand control

Institutional Sale: REIT/family office buyout at brand premium

Portfolio Roll-Up: Multiple city deployments → packaged sale at EBITDA multiple

7. WHY VELOCITY IS DIFFERENT

Scarcity Economics: Engineered supply limits, zoning barriers, cultural demand surge

AI Edge: Predictive demand modeling, AI-driven pricing (+14% premium over comps), and automated investor dashboards

Brand Moat: Lifestyle-driven, identity assets → more than storage, it's status

Downside Protection: Conservative leverage, pref returns, subordinated developer capital

Key Investor Takeaway

Velocity Performance Alliance offers investors institutional-grade risk protection with venture-style upside. By structuring a disciplined capital stack, delivering stabilized EBITDA, and building a replicable branded platform, VPA positions LPs for both strong near-term cash returns and long-term exit multiples.

OPTIONAL ADD-ONS

- Watermarked version for soft-circulation
- DocSend-enabled interactive PDF
- Integrated digital signature field
- AI-powered video walkthrough embedded via QR
  - **1. Velocity Lead Magnet Funnel (VLMF)**
    Auto-launch tool for generating location-based leads
    Platform: HighLevel or Webflow + GPT copy engine
    Prompt: "Generate landing page headline + video sales letter (VSL) for car collectors in Scottsdale, AZ seeking secure equity storage."

- **2. Deal Evaluator GPT**
  Input address, asking price, storage unit count
  Output: projected cash-on-cash, equity split scenarios, IRR in 3 exit models
- **3. Auto-Condo AI Negotiator™**
  Generates counter-offers based on seller language, emails, and listing tone
- **4. Licensee Performance Dashboard**
  Tracks per-territory revenue, lead velocity, close ratio, and attrition risk

D. PITCH-SELL-CLOSE (PSC) SCRIPT

Pitch:

"Let me ask you a serious question: What if your car could pay for itself while sitting in a temperature-controlled, AI-secured vault, and you owned the vault?"

Sell:

"We had a client in Miami who was a Ferrari collector and never trusted the market. One unit became five. Now he's generating more income from his car condos than his brokerage account."

Close:

"There are two types of people: those who invest when they get it, and those who wait until the market validates it. The first group builds wealth. The second pays retail. If you see yourself in the first group, let's structure a deal right now."

D1. CASE STUDY: VELOCITY ROW SUITES — SCOTTSDALE, AZ

How AI validated market entry, unit mix, and scarcity-driven monetization

In the world of specialized real estate, the Scottsdale project stands as a benchmark for how precision meets profit. A $46 million development that aligns perfectly with JT Foxx's real estate execution model: data-driven market entry, speed-to-sale, and compounding value through strategic upgrades and investor psychology.

It was not just about building storage. It was about creating a branded experience with engineered scarcity and built-in upsell pathways. Here's how it was done and how you replicate it.

D2. AI-MAPPED MARKET SELECTION: WHY SCOTTSDALE?

AI Inputs Used:

- Affluence Density & Vehicle Ownership: Machine learning tools cross-referenced IRS income data, Department of Motor Vehicles (DMV) luxury vehicle registrations, and RV ownership ratios. Scottsdale's North/East zip codes showed more than 3.2 times the national average in collector vehicle ownership.
- Zoning Compliance Mapping: AI scraped local zoning regulations across Maricopa County to find light industrial overlays with minimal Not In My Backyard (NIMBY) pushback and streamlined entitlement processes.
- Real-Time Sentiment: Google Trends and Meta audience analytics identified a spike in "garage condo," "RV storage," and "car collector storage" queries from affluent males aged 40–65.
- Outcome: AI validated that Scottsdale wasn't just viable; it was optimal. High-income demographics, car culture obsession, and low existing competition created the perfect launchpad.

D3. AI-CALIBRATED UNIT COUNT & SALES VELOCITY

Target: 64 Units, Built in 3 Phases

- AI-Driven Demand Modeling:
  - AI regression tools used data from 5 comp projects (AZ, FL, IL) to forecast Scottsdale's absorption rate.

- Optimal unit count was determined using a Monte Carlo simulation to avoid overbuild risk while maintaining price escalation upside.
- Data-Backed Build Strategy:
  - 32 units prepped for sale by Month 14, matching forecasted local demand curve.
  - Each phase included a 5–7% price escalation, modeled using behavioral economics algorithms to drive FOMO and compress decision cycles.
- Key Insight: Building 64 units instead of 80 or 100 kept risk exposure low, preserved perceived scarcity, and maximized ROI through phased premiums.

D4. MONETIZATION STRATEGY POWERED BY AI
- Shell Sale Price (Base): $725,000
- Premium Packages: Up to $200,000 per unit
- Total Gross Revenue: $62.93M
- Total Cost Basis: $46.01M
- Total Modeled Profit: ~$16.0M

AI modeled three core revenue layers:
- Base Shell Sales
- Premium "Event-Ready" & "Security" Upgrades (modeled at 70% gross margin)
- HOA Cash Flow Valuation as Yield Asset

Scarcity Triggers Modeled:
- Limited-release packages (Monaco, Dubai Editions) tested via ad copy AI before launch.
- Tokenized structure embedded liquidity events within 24–36 months, appealing to both speculative and yield-focused capital.

D5. SALES STRATEGY: AI-FUELED PIPELINE ACTIVATION

Pipeline Secured Before Groundbreaking:
- Over 20 soft commitments secured using behavioral AI outreach tools.
- Three-broker team managed a CRM funnel of 150+ pre-qualified prospects with predictive scoring applied to buyer behavior.
- AI optimized messaging cadence, timing, and content by segmenting buyers into emotional versus utilitarian profiles—then personalized outreach accordingly.

D6. LESSONS FOR THE DEVELOPER

What JT Foxx Would Tell You:
- Don't just build. Monetize from day one by engineering urgency.
- Use AI not to replace instinct, but to remove risk and accelerate decisions.
- Position premium upgrades as emotional lifestyle anchors—not options.
- Lock in margin at the design phase, and stack profit in layers, not just at exit.

E. SAMPLE PRO FORMA FROM AI

PROJECT OVERVIEW
Velocity Row Suites | Phoenix metro area, AZ
A Legacy-Class Investment in Luxury Automotive Lifestyle

Velocity Row Suites is to be a first-in-class luxury automotive condominium development located in Phoenix metro area, AZ; the epicenter of Arizona's car culture. Designed to attract high-net-worth collectors, business owners, and lifestyle buyers, this development integrates customizable vehicle vaults with concierge-level amenities and curated high-end events.

The project will unfold in three distinct phases: creating recurring momentum, capital velocity, and multiple monetization windows.

- THE VISION
  Not just a garage; a lifestyle asset. Velocity Row is building a brand, not just a building:
- Phase 1: 64-unit vehicle vaults (128,000 SF)
- Phase 2–3: Amenitized expansions including concierge services, private lounges, exclusive event access, and resale arbitrage through HOA-aligned value capture.

INVESTOR PRO FORMA HIGHLIGHTS - PHOENIX AZ METRO AREA

PROJECTED PROJECT COST
- Velocity Row Suites – Scottsdale, AZ
- Total Development Cost (TDC): $46M
- Investor Equity Raise (Cap): $6.9M
- Developer Equity Committed: $2.3M+
- Capital Stack Security:
- Senior Debt (60% LTC): $27.6M (Fixed, Interest-Only)
- Preferred Equity: $9.2M @ 8% secured
- Investor Equity: 70/30 split post-8% preferred
- Downside Protection: 2% minimum IRR return clause before promote

PROJECTED RETURNS
- Total Gross Revenue: $62.93M
- Target IRR: 22.1% – 26.4%
- Upside Driver: Inclusion in national expansion play increases buyer demand (institutional and high-net-worth private capital)
- Projected Unit ROI: $312/sq ft build cost vs. $675/sq ft exit value
- Exit Route: Either refinance with equity roll or full asset disposition
- Exit Valuation: Framed not just on cap rate, but on perceived scarcity + brand utility
- Net Profit: $15.51M
- Equity Multiple (Target): 1.68x – 2.5x (time-based)
- Stabilized EBITDA: $2.56M
- Valuation @ 6–8x EBITDA: $15.4M – $20.5M

WHY NOW? WHY THIS?
- Scarcity-Driven Demand: The Phoenix metro area is a magnet for automotive elites supporting premium secure vehicle vaults with integrated lifestyle? Zero direct comps.
- Early Mover Advantage: First to market with phased buildout. Investors get in below the curve, not at the peak.
- Margin Stack Engineered for Exit: 32.7% project margin; multiple ROI layers including lifestyle upsells, HOA yield, and recap triggers.

- Risk-Mitigated Structure: Equity is protected through preferred returns, backend participation, and enforceable downside clauses.

This is not just a real estate play; it's an access play. Access to an ultra-high-net-worth network, to the Phoenix metro area's elite buyer class, and to the kind of legacy asset that builds not just cash flow, but credibility. Velocity Row isn't selling space, it's selling identity. This project has a $6.9M cap. Once closed, no future equity rounds offered. Investors must commit before vertical build begins.

CAPITAL STACK OVERVIEW
Structured for Protection and Performance

Capital Stack Summary
This structure is designed to prioritize investor security, optimize tax efficiency, and align returns with execution milestones.

Capital Layer	Amount	Position	Terms
Developer Equity	$2.3M	Subordinated (5% of TDC)	Risk capital, last paid
Investor LP Equity	$6.9M	Common Equity (15% of TDC).	8–10% Preferred Return, 70/30 split
GP Co-Invest (optional)	Variable	Pro-rata with LP (if elected)	Aligned at LP terms
Tokenized Tranche (TBD)	Optional	Smart contract yield layer	For future liquidity partners
Preferred	$9.2M	Junior Secured 2nd Position	(20% LTC) 8% Fixed, common priority over Equity
Senior Debt	$27.6M	1st Position (60% LTC).	Negotiated variable -rate, interest-only
Total Project	$46M		

Cash Flow Waterfall (Simplified):
- Return of LP Capital
- 2% IRR Minimum to LPs
- 8–10% Preferred Return (annualized)
- Profit Split: 70% LP / 30% Sponsor

This structure ensures investors are paid first and in full before any promote is triggered. The 2% IRR minimum creates downside insulation even in delayed monetization scenarios.

IRR Sensitivity – Modeled Scenarios

Exit Timeline	Projected IRR	Equity Multiple
24-Month Monetization	24%+	2.5x+
36-Month Recapitalization	21%	2.0x–2.3x
48-Month Exit/REIT Sale	17%	1.8x+

These returns are net of all fees and structured to scale with timing flexibility — allowing repositioning based on market conditions or brand platform expansion.

CAPITAL STRUCTURE
Strategic Protection. Aligned Upside. Flexible Institutional Exit.

The capital structure is built to prioritize investor protection while maximizing risk-adjusted returns. Each layer is engineered with sequencing, preference alignment, and contractual downside coverage to create both security and scalability.

1. Senior Debt (Institutional or Private Lenders)
   - Role: Foundation of leverage; lowest risk position.
   - Protection Features:
   - First lien on property + assignment of contracts
   - Interest reserves fully capitalized at closing
   - Conservative loan-to-cost (LTC) ratios (≤ 65%)
   - Investor Advantage: Secured exposure, predictable coupon
   - Exit Flexibility: Refinancing or sale proceeds used to retire debt first

2. Mezzanine / Preferred Equity
   - Role: Bridge layer between senior debt and common equity.
   - Protection Features:
   - Priority distributions before common equity
   - Target IRR with capped upside for predictability
   - Intercreditor agreements to protect against senior lender foreclosure
   - Investor Advantage: Higher yield than debt, downside insulation before common equity
   - Exit Flexibility: Convertible features allowing takeout into equity on recap

3. Common Equity (Velocity + Strategic LPs)
   - Role: Alignment layer driving value creation and upside capture.
   - Protection Features:
   - Equity waterfall ensures preferred equity fully satisfied first
   - GP (Velocity) co-investment ensures skin-in-the-game discipline
   - Restricted distributions until construction milestones + sales thresholds are met
   - Investor Advantage: Institutional-level upside through profit participation
   - Exit Flexibility: Multiple exit paths (condo sellout, portfolio recap, REIT bulk sale)

4. GP Promote / Sponsor Equity (Velocity Performance Alliance)
Role: Performance incentive tied only to exceeding investor return hurdles.
   - Protection Features:
   - Promote earned only after LPs receive capital + pref return
   - Clawback provisions for investor-first recovery
   - Performance-based vesting tied to sales velocity + NOI thresholds
   - Investor Advantage: Sponsor incentivized to overperform, not just deliver minimums
   - Exit Flexibility: Promote crystallizes only on exit liquidity event

Sequencing & Distribution Framework (Waterfall)
   - Return of capital to all investors
   - Preferred return distributed to Pref Equity LPs (8–10% hurdle typical)
   - Remaining profits split:
   - 70–80% to LPs (until defined IRR threshold)
   - 20–30% to GP Promote upon exceeding thresholds
   - Contractual Downside Coverage
   - Completion guarantees from GC and Velocity principals
   - Cost overrun reserves pre-funded into construction escrow
   - Key-man insurance protecting against principal disruption
   - Investor step-in rights if sponsor fails to meet reporting or capital obligations

AI-Enhanced Capital Oversight
- Predictive waterfall modeling under stress-test scenarios (interest rate hikes, absorption delays)
- Automated drawdown tracking against budget variances
- Real-time IRR/Equity Multiple dashboards for LP visibility
- AI-triggered compliance locks on distributions until pref + return hurdles are met

Capital Stack Layers (Illustrative Diagram)

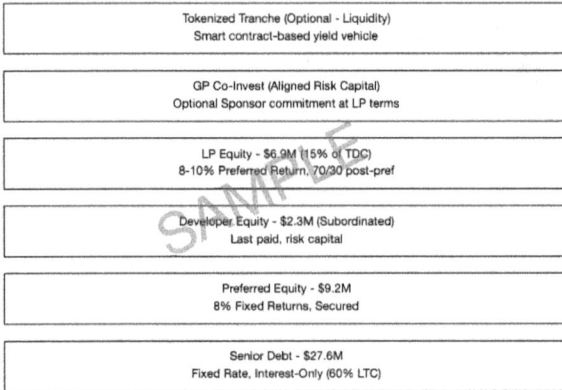

Tokenized Tranche (Optional - Liquidity) Smart contract-based yield vehicle
GP Co-Invest (Aligned Risk Capital) Optional Sponsor commitment at LP terms
LP Equity - $6.9M (15% of TDC) 8-10% Preferred Return, 70/30 post-pref
Developer Equity - $2.3M (Subordinated) Last paid, risk capital
Preferred Equity - $9.2M 8% Fixed Returns, Secured
Senior Debt - $27.6M Fixed Rate, Interest-Only (60% LTC)

Tokenized Tranche (Optional - Liquidity)
Smart contract-based yield vehicle

GP Co-Invest (Aligned Risk Capital)
Optional Sponsor commitment at LP terms

LP Equity - $6.9M (15% of TDC)
8-10% Preferred Return, 70/30 post-pref

Developer Equity - $2.3M (Subordinated)
Last paid, risk capital

Preferred Equity - $9.2M
8% Fixed Returns, Secured

Senior Debt - $27.6M
Fixed Rate, Interest-Only (60% LTC)

Sample Cash Waterfall Structure
- Return of LP Capital
- 2% Minimum IRR Clause (Downside Protection)
- 8–10% Preferred Return (Secured)
- 70/30 Split of Excess Cash Flows
- 70% to LP investors
- 30% to sponsor/developer (promote)
- Optional Token Tranche (future-enabled liquidity layer)

Note: No promote or profit participation is triggered until LPs receive full capital return + minimum IRR.

IRR MODELING BY EXIT TIMING

Modeled returns across three strategic exit windows:

Exit Timeline	IRR (Net)	Equity Multiple
24-Month Monetization	24%+	2.5x+
36-Month Recap	21%	~2.3x
48-Month Hold	17%	~1.8x

Sensitivities based on timing, pricing and revenue uplift, and backend equity structure.
Returns reflect backend participation after preferred returns and minimum IRR protection.
Velocity's stack is not just layered — it's locked for protection, geared for performance, and expandable as a global platform.

SAMPLE PROJECT FINANCIALS
Velocity Row Suites – Phase 1 | Phoenix metro area, AZ

Key Financial Metrics

Category	Detail
Construction Cost	$273/SF (Total: $46M TDC)
Unit Pricing	Base: $725K / Premium: $800K–$940K
Total Gross Revenue	$63M (64 units + upsells + HOA yield)
Modeled Net Profit	~$16M on cost basis
EBITDA (Stabilized Year)	$2.56M
Valuation (6x–8x EBITDA)	$15.4M – $20.5M
Profit Margin	32.7%
Hold Period	24–48 months (performance-based)
Revenue Layers	Shell sales, event/lifestyle upsells, HOA cap yield

Revenue Model Breakdown
Shell Sales: $46.4M
Lifestyle/Event Upsells: $8.0M
Security Packages: $4.8M
HOA Capital Yield (modeled exit): $3.73M
All revenue sources are integrated into a three-phase tiered inventory release, optimizing pricing escalation, urgency, and buyer velocity.
Sponsor Investment Structure & Rolled Equity:
Velocity Row's GP is contributing $2.3M of subordinated developer equity — but its impact is significantly magnified due to one core reason:
Profits from early phase monetization are being rolled forward into future phases.

This means the GP's position is not only performance-aligned but also internally capitalized, reducing dilution, accelerating backend upside, and preserving equity control without over-raising.

By reinvesting gains from early unit sales and upsells, the GP effectively compounds sponsor equity without requiring significant fresh capital, demonstrating confidence in execution and discipline in capitalization.

This is not speculative. This is phased, margin-stacked, and built to compound.

INVESTOR DATA ROOM ACCESS
Velocity Performance Alliance | Digital Capital Command Center

Access Portal
All approved investors will receive access to the Velocity Performance Alliance Digital Data Room; a secure, centralized platform for reviewing, executing, and tracking your investment.

URL: www.VelocityInvestorAccess.com
QR Code: Will be generated in the near future
Secure Login: Provided upon verification of accredited status

Inside the Portal
Private Placement Memorandum (PPM)
Investor Deck & Pro Forma
Subscription Agreement & DocuSign Access
Smart CRM Dashboard
Allocation status
Capital call schedule
Distribution tracking
Exit notification alerts
GPT-Powered Q&A Interface for instant deal-specific questions

Access & Support
To request secure credentials or schedule a live walk-through:

Investor Relations Team
Email: invest@velocityperformancealliance.com
Phone: +1 (512) 942-5969
Portal Support: login@velocityinvestoraccess.com

Your capital deserves transparency, speed, and control. This platform delivers all three.

MONETIZABLE EXIT METRICS

Metric	Velocity Benchmarks
Time-to-Sell Premium Units	37% faster than market comps
AI Price Optimization Delta	+14.8% above broker CMA
Syndicated Investor Yield	≥ 20% net of fees
Brand Licensing Multiplier	1.5x traditional hard asset exit

NEXT STEPS + CONTACT
Velocity Performance Alliance | Final Stage Access

Join the Syndicate
- Minimum Commitment: $250,000
- Investor Status: Accredited only
- Raise Status: Active, limited to $6.9M
- Close Date: Prior to vertical build commencement

This is a closed-round raise. Once capital is allocated, no future equity participation will be offered.

Engage Directly
Schedule a Private Briefing
Book Virtual Data Room Walkthrough
Our team is available to customize your onboarding based on capital tranche, hold preference, and legal structure alignment.

Contact Details
Velocity Capital Team
Email: capital@vxxxxxxxxxxxxxxxxx.com
Phone: +1 (555) 555-5555
Portal: www.xxxxxxxxxxxxxxxx.com

Legal & Compliance Counsel
For accredited investor verification, entity structuring, and documentation:
Email: compliance@velocityperformancealliance.com

This is the final stage before capital locks. You're either in before the vertical goes up — or you're out.

F. AI-Generated Design of Auto Condominiums

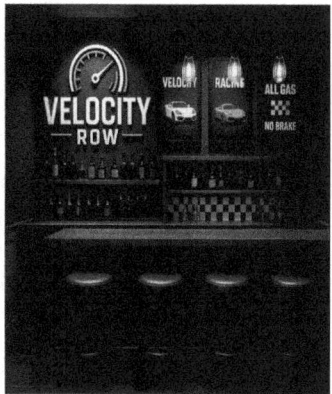

CONCLUSION

This book began with a simple premise: in real estate, you don't win by being the biggest—you win by being the smartest. Through each chapter, we've shown that artificial intelligence is not a futuristic luxury; it is the decisive weapon for developers, investors, and entrepreneurs who want to out-think, out-build, and out-profit their competition.

Luxury auto condominiums and other niche developments are more than physical assets. They are engineered movements—where identity, scarcity, and brand equity converge to create trophy assets that command premium margins and global demand. With AI, you gain precision in site selection, speed in execution, and influence in capital markets. With brand strategy, you transform square footage into emotional value that investors and buyers can't ignore.

The message is clear: this is not about theory—it's about execution intelligence. The frameworks in these pages give you the unfair advantage to frame capital raises, dominate zoning, engineer wealth triangles, and pre-sell projects before breaking ground.

As you step forward, remember—your competition will cling to old models of spreadsheets, instinct, and luck. Your edge is clarity, control, and conviction. The developers and entrepreneurs who embrace AI-driven execution will not just build projects; they will build categories, legacies, and generational wealth.

Throughout this book, we've explored the strategies, systems, and frameworks that separate surviving developers from category-defining entrepreneurs. The lesson is clear: in today's market, victory does not belong to the biggest—it belongs to the smartest.

Artificial intelligence, strategic branding, and capital structuring are not theories; they are execution weapons.

When deployed with precision, they allow you to:

Raise capital faster and on better terms by framing certainty, emotion, and exclusivity.

Develop with speed and clarity by controlling land, zoning, and execution before competitors even see an opportunity.

Build the Real Estate Wealth Triangle—cash flow, appreciation, and scalability reinforcing one another into generational wealth.

Brand developments as identity assets, not just buildings, so that buyers and investors are pre-sold before the first shovel hits the ground.

This is execution intelligence—the discipline of turning uncertainty into clarity, instinct into precision, and projects into platforms. It is the ultimate arbitrage, available only to those willing to think differently and act decisively.

Epilogue

As we close these pages, I want to leave you with a truth that transcends blueprints, spreadsheets, and zoning maps: success in real estate—or in any business—is not about buildings. It's about people, psychology, and the choices we make to seize opportunity before the rest of the world recognizes it.

I began this journey with projects that barely survived. What turned the tide wasn't luck, or even timing—it was a complete rewiring of how I thought, acted, and executed. Mentorship, frameworks, and the integration of AI gave me what so many developers lack: clarity, speed, and certainty. Where others hesitated, I acted. Where others built for today, I built for legacy.

The projects in this book—auto condominiums, niche developments, branded identities—are more than investments. They are identity assets. They represent the ability to take someone's passion and turn it into permanence, to transform status into an asset class, and to engineer desire into long-term wealth.

The future will not be kind to developers who cling to instinct, tradition, or hope. It belongs to those who fuse human wisdom with machine precision, who treat branding as a weapon, and who understand that scarcity and story will always command more than square footage ever could.

If you remember nothing else, remember this: execution intelligence beats size, legacy, and luck. Every great fortune starts with the courage to act differently. So take these strategies, test them, refine them, and make them your own. Build movements, not projects. Create categories, not comps. Think in decades, not deals.

And most importantly—leave behind more than returns. Leave behind assets that become legacies, brands that endure beyond your lifetime, and a story that proves you did not just build real estate—you built something that mattered. But beyond the numbers and strategies lies something more enduring. Real estate, at its core, is not about steel, glass, and land. It is about people. It is about psychology, story, and the courage to lead when others cling to tradition.

I began with projects that were struggling to survive. What transformed everything wasn't timing or luck—it was a complete rewiring of how I thought and acted. The combination of mentorship, proven systems, and AI-driven clarity turned survival into growth, and growth into scale.

Auto condominiums and other niche assets are not "garages" or "units." They are sanctuaries of identity, engineered scarcity, and branded ecosystems that buyers and investors can't ignore. When you see that truth, you stop competing on comps and start setting the standard. You stop following markets and start creating them.

The developers and entrepreneurs who embrace this path will not just build projects; they will build movements. They will own categories. They will leave behind legacies measured not only in IRR and cash flow, but in the stories, brands, and identities they engineered into permanence.

So, as you close this book, remember: execution intelligence beats size, legacy, and luck—every time.

Build movements, not projects. Create categories, not comps. Think in decades, not deals. Use AI not as a buzzword, but as your competitive weapon.

And above all—don't just build real estate. Build something that matters. Something that endures. Something that proves you were not in the business of creating square footage, but in the business of creating legacies.

The blueprint is now in your hands. The only question left is: will you execute?

CLOSING STATEMENT

AI STRATEGIES FOR AUTO CONDOMINIUM DEVELOPMENTS: PROFITABLE EXECUTION IN THE FUTURE OF SPECIALIZED REAL ESTATE

In an era defined by speed, scarcity, and precision, luxury auto condominiums are no longer just storage—they're statements, ecosystems, and investment-grade lifestyle assets. As the global elite seek rare, branded real estate that aligns with their passions, the market will reward those who build faster, smarter, and more deliberately.

The strategies in this book are not just speculative—they are executable. They don't require hundreds of employees or nine-figure budgets. They require focus, tools, positioning, and AI leverage.

You now have in your hands a blueprint to:
- Replace inefficiency with intelligence
- Convert AI into investor trust and margin protection
- Use behavioral data to attract high-net-worth buyers
- Scale regionally without bloated capital commitments
- And, ultimately, exit not just with profit—but with power.

The future of real estate belongs to specialists, not generalists. To creators of branded, data-rich, high-emotion assets. To developers who are part entrepreneur, part technologist, part storyteller.

Whether you are a first-time developer or seasoned professional, the truth is clear:

AI is not optional. It is the new foundation.

This book is your invitation to lead the next wave. Not just in real estate, but in the business of rare value.

The race has already started. Those who wait will fund those who act.

Now go build what others will wish they had the courage to begin.

Glossary

Active Income— Direct earnings from sales or services in a real estate venture (e.g., pre-sales, concierge packages). Part of the Real Estate Wealth Triangle.

AI Advantage— The integration of artificial intelligence into every phase of development to maximize precision, reduce risk, and accelerate execution.

AI Branding Strategies— Use of AI to engineer perceived value through naming, storytelling, exclusivity, and emotional identity—leading to pre-sold projects before ground is broken.

AI Exit Mapping— The process of using predictive analytics to model likely buyers (REITs, PE funds, family offices) and exit timing, improving certainty for investors.

AI-Driven Site Selection— The use of geospatial, demographic, and psychographic data to pinpoint ideal parcels before they hit the market.

Asset Appreciation— The increase in value of both the underlying land/structure and the branded IP, producing premium exit multiples.

Auto Condominiums— Luxury, climate-controlled garage units designed as sanctuaries for UHNW collectors. Sold as identity assets, not storage.

Behavioral Buyer Targeting— AI-driven analysis of buyer behavior to identify emotional triggers and micro-signals that lead to conversion.

BIM (Building Information Modeling)— A dynamic, 3D data-rich model that enables design optimization, clash detection, and investor visualization.

Bill of Materials (BOM)— Detailed inventory of all construction components. Used to manage costs, protect margins, and negotiate with vendors.

Brand Equity— The intangible value of a project's brand, which often exceeds the physical real estate in long-term worth.

Capital Stack— The layered financing of a project (debt, preferred equity, common equity, tokenized securities).

Category Ownership— Positioning strategy where a developer creates a new asset class and sets market value rather than competing on comps.

Concierge Upsells— Revenue streams from add-on services such as event hosting, detailing bays, or custom fit-outs.

Cost Engineering— Using AI to reverse-engineer projects for profitability by optimizing layouts, materials, and schedules.

Dynamic Investor Tiers— Framing investment rounds as Founder, Premier, or Core tiers—each with distinct incentives and prestige levels.

Dynamic Pricing Models— AI-driven real-time and tiered pricing systems based on demand, scarcity, and urgency.

Emotional Equity— Investor alignment built on prestige, legacy, and trust rather than spreadsheets.

Execution Intelligence— The disciplined application of AI-powered systems to eliminate guesswork and accelerate profitable outcomes.

Framing Stack (JT Foxx)— Capital-raising method that anchors with certainty, elevates investor identity, and sequences logic after emotion.

Franchise Model— Scaling strategy where a developer licenses brand and systems to partners while retaining oversight and a revenue share.

Generative Design— AI process that rapidly tests thousands of design iterations for cost, efficiency, and luxury appeal.

HOA Revenue— Recurring income from homeowners' association fees, event hosting, and amenities.

Identity Real Estate— Assets purchased for belonging and status rather than utility—e.g., auto condos or branded clubs.

Investor Machine— Perpetual AI-driven capital raising system that automates outreach, nurturing, and conversion.

JT Foxx AI Sales System (PSC)— Sales methodology of Pitch–Sell–Close enhanced with AI to convert leads within 21 days.

Licensing— The sale of brand and operational systems for replication in new geographies, creating ongoing revenue.

Lifestyle Branding— Marketing projects as exclusive lifestyle clubs rather than real estate, elevating perceived value.

Macro Surge— The broader wealth and collectible trends that create tailwinds for niche asset demand.

Niche Real Estate— Specialized developments (auto condos, medical retreats, branded communities) designed for premium margins.

Operational Automation— AI-enabled back-end management of contracts, onboarding, concierge, and resale operations.

Perceived Scarcity— Engineered exclusivity through invite-only phases, countdowns, and founder groups.

Predictive Analytics— AI use to forecast demand, construction costs, buyer churn, and exit timing.

Pre-Sell Model— Strategy of selling units before construction begins, funded by deposits and engineered scarcity.

PSC Scripts— AI-tailored conversational flows that guide buyers through Pitch, Sell, and Close.

Real Estate Wealth Triangle— Framework aligning Cash Flow, Asset Appreciation, and Speed to Scale into a compounding wealth engine.

Replication Engine— System for duplicating successful developments across cities using AI-driven design, marketing, and investor systems.

Risk Mitigation— AI-based strategies to identify zoning risk, budget overruns, or exit delays before they occur.

Scarcity Economics— Deliberate limitation of supply to increase pricing power and urgency.

Smart Entitlements— AI-assisted permit and zoning tracking tools that reduce delays and flag risks.

Tokenization— Blockchain-based issuance of fractionalized securities that increase liquidity and expand the investor pool.

Velocity Performance Alliance— Proprietary system combining AI, branding, and structured capital to deliver scalable, high-margin niche developments.

Wealth Licensing— The monetization of intellectual property and brand equity by licensing models globally.

Zoning Domination— Securing favorable zoning by mapping political influence, aligning with municipal branding, and controlling the public narrative.

INDEX

ABOUT THE AUTHOR
WILLIAM BÖLL

A Founder, Board Member, and Chief Operating Officer at Oasis Design & Development Corporation, a green and eco business development and project delivery company whose real estate development products are Velocity Prestigious Auto Residences and Velocity Performance Alliance, a NASCAR-themed auto condominium.

In addition, he is a Co-Founder of nVolve Technologies, a building "technology" company that focuses exclusively on developing new construction material technology and processes that allow us to build better, energy-efficient residential, commercial, and retail buildings very cost-effectively. William works directly in the areas of leadership, business development, innovation, operational analysis, negotiation, strategic planning, and green and sustainable building materials and technologies. He is recognized for establishing, growing, and managing top-performing teams. With over 25 years of experience, William has completed over $4 billion in worldwide projects encompassing 2.2 million SF of builds. His team has completed over $9 billion in worldwide projects encompassing 30 million SF of builds.

He is skilled in recruiting, developing, and leading channel expansion, business development, product management, operations, and logistics teams to consistently exceed goals across constantly evolving business and market environments. William processes strong innovation, transformation, and growth leadership with proven success in introducing standards and operational best practices that have reduced operating expenses, improved quality, and increased production and overall market share.

His experience spans global technology as well as

Mergers & Acquisitions. Clients include Johnson & Johnson, Lufthansa, Royal Dutch Shell, Allstate Insurance, Bank of Montreal, CIBC, and Toronto Dominion Bank.

Outside of the corporate world he is a retired supervisor in multiple professional hockey leagues.

Coming Soon

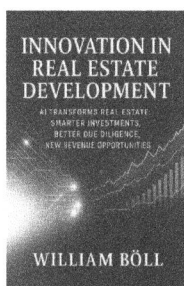

AI Innovation in Real Estate:
Discover How Artificial Intelligence
is Transforming Real Estate
Investment and Development into
Revenue Optimization

INNOVATION IN
REAL ESTATE
DEVELOPMENT

WILLIAM BÖLL

Real estate is one of the world's most established industries—yet it is also one of the ripest for disruption. In AI Innovation in Real Estate, readers are introduced to a robust roadmap for how artificial intelligence is transforming the way we invest, develop, and manage the built environment.

The book begins by laying the foundations of the AI revolution in real estate. It explores how machine learning, deep learning, natural language processing, computer vision, and predictive analytics unlock new levels of efficiency, accuracy, and foresight. With data as the backbone, readers learn how to source, clean, and apply real estate information in an ethical and effective manner.

From there, the book moves into real-world applications. In investment, AI helps identify undervalued opportunities, automates due diligence, and rebalances portfolios in real-time. In development, generative design accelerates feasibility studies and produces optimized floor plans tailored to market demand, while predictive models enhance sustainability and energy efficiency. On the construction site, AI mitigates risk, monitors progress with drone and camera feeds, and streamlines supply chains, thus reducing delays and cost overruns.

The scope then widens to cities, where AI is driving smarter infrastructure planning, mobility systems, and sustainable growth. Case studies demonstrate how forward-thinking developers, funds, and municipalities are already leveraging AI to gain a competitive edge.

The book concludes by addressing the challenges—bias in algorithms, data privacy, regulation—and offering a clear strategy for adoption. By combining practical insights with future-focused vision, AI Innovation in Real Estate equips investors, developers, and urban leaders with the tools to thrive in an industry being reshaped before our eyes.

AI Innovation in Real Estate is more than a book; it is a blueprint for the future of the real estate industry.